Greening of Industry Networks Studies

Volume 8

Series Editors

Diego A. Vazquez-Brust, University of Portsmouth, Portsmouth, Hampshire, UK
Joseph Sarkis, Worcester Polytechnic Institute, School of Business, Worcester, MA, USA

The Series aims to improve our understanding of how shifts in industrial regimes, trade, and technology are creating significant environmental and social impacts and inequities around the world; but also opportunities for sustainable economic growth. The series will endeavor to develop knowledge and transform practice to accelerate a paradigm change toward a sustainable society across disciplines, geography, and sectors. As such, the series will be an integral part of GIN-3D (Greening of Industry Network Third Decade) strategy. The trajectory began a couple decades ago and is still evolving. Books in the series will help document this trajectory as well as point it in the appropriate direction, thus we welcome all contributions from scholars– GIN members and non-GIN members- working to steer the transition to a better, green and just future.

More information about this series at http://www.springer.com/series/10444

Jafar Rezaei
Editor

Strategic Decision Making for Sustainable Management of Industrial Networks

Springer

Editor
Jafar Rezaei
Delft University of Technology
Delft, The Netherlands

ISSN 2543-0246 ISSN 2543-0254 (electronic)
Greening of Industry Networks Studies
ISBN 978-3-030-55384-5 ISBN 978-3-030-55385-2 (eBook)
https://doi.org/10.1007/978-3-030-55385-2

This Springer imprint is published by the registered company Springer Nature Switzerland AG
The registered company address is: Gewerbestrasse 11, 6330 Cham, Switzerland

Acknowledgements

I would like to thank all the contributors of the book for their excellent contribution and patience. Special thanks go to the team at Springer for their help and guidance in the completion of the book.

Strategic Decision-Making for Sustainable Management of Industrial Networks: An Introduction

Sustainability

Sustainability is a new field of science which tries to understand, analyze, and improve "the fundamental character of interactions between nature and society" (Kates et al. 2001). Sustainability cares about meeting the needs of the present and future generations of humans. There exist several other concepts which although have some linkages with sustainability, they are no synonyms. These include circular economy, environmental science, and environmental protection, among others. While sustainability cares about ecological, economic, and social objectives, environmental protection and environmental science have more focus on ecological objectives, and circular economy (CE) has a focus on ecological and economic objectives (Sauvé et al. 2016). CE is a new business model with a focus on efficiency in resources through waste prevention, exchange, reuse, and recycling patterns (Geng et al. 2016). While some chapters of the book have a more focus on ecological or environmental sustainability, in some chapters the focus extends to social and economic dimensions of sustainability based on the triple-bottom-line definition of sustainability. Having a long-term perspective, one might argue that destroying the nature could ultimately come at the cost of destroying the society, which is why in research on sustainability, like in this book, environmental sustainability has a primary role.

What we, as individuals or organizations, decide affects the nature that we are part of. This implies that if we decide to destroy (preserve) the nature, we ultimately destroy (preserve) ourselves. So, our decisions matter! This is why in this book, we try to understand the role of decision-making for sustainable management. The book has a focus on strategic decisions in industrial networks. Today, many organizations take big steps towards sustainability. They try to use cleaner production technologies, use renewable energies, and improve health and safety issues in their organizations and the products and services they offer. These involve several important strategic and managerial decisions, which is the topic of this book. The main decision-making areas that the book covers include renewable energy, green innovation, logistics in a circular economy, interconnected logistics, sustainable green freight transport, agri-food in circular economy, territorial transformation, sustainable supplier segmentation, and information overload and health. The aim of the book is not to focus on one particular industry or a particular topic. It is rather to see how strategic and managerial decisions are taken in different industries and topics around sustainability.

Managerial and Strategic Decision-Making

"Managers are invested with the task of making decisions which routinely affect the value and viability of firms" (Rowe et al. 1984). Decision-making is generally defined as a cognitive process which results in evaluating and selecting one or more alternatives from among a set of alternatives. The decisions which are made by managers not only could affect the value and viability of their firms but also could affect the value and viability of their environment. Each and every decision made by managers could impact the economic, environmental, and social dimensions of their firm and their environment. For instance, designing package for food products could impact the profitability of the company, the health of consumers of the product as well as the environment depending on whether the material used in the package can be recycled or not (Rezaei et al. 2019).

Over the past decades formal theories of decision-making such as expected utility theory and game theory have been recognized as models of rational decision-making in organizations (Simon 1979). Rational decision-making theories assume that people/organizations optimize their goals (maximizing profit, minimizing costs). While very useful in many circumstances, in some other circumstances, these theories are not able to fully represent the way people/organizations make decisions. Behavioral decision-making theories (Slovic et al. 1977; Kahneman and Tversky 2013) have tried to model decision-making process closer to its reality. Sustainability is a multi-dimension concept. This implies that making decision in sustainability is difficult. It contains economic, social, environmental dimensions which cover many soft and hard aspects, from cultural aspects to chemical formulation of the materials used in production. One should analyze many factors which form sustainability and at the same time investigate the impact of the decisions. This is why many researchers in this field investigate one or some dimensions of sustainability and not its whole. It is about a trade-off. Conducting deep analysis necessitates narrow focusing while getting a holistic comprehension comes at the cost of details. We need both.

In this book, each chapter will discuss one or several decision-making problems with potential impact on sustainability. While some chapters are deep into a particular dimension of sustainability in a particular decision-making context, some provide a rather big picture. Perhaps, as one direct consequence, the methods used are also very diverse from optimization to multi-criteria decision-making methods to descriptive methods.

Overview of the Book

There are nine chapters in this book excluding this introductory chapter. The book covers the work of several international contributors from different fields of study who share their research in sustainability in different industrial networks. The two main themes of the book (sustainability and strategic decision-making) are too broad to provide an exhaustive cover of all relevant topics. The book is rather providing a sample of such strategic and managerial decision-making in some particular contexts. Although there is some overlap among the chapters, care was taken to make each chapter standing alone. Below a short summary of each chapter is given.

Chapter 1 (**Beames, Claassen, Akkerman**) discusses the logistics challenges in the context of circular economy (CE) with a particular focus on circular business innovations, servitization, and open business models. Three key strategic decision-making problems for businesses for the transition towards circular business models are discussed: (i) the extent to which the logistics network is centralized, (ii) the extent to which the product is servitized, and (iii) the extent to which logistics services are coordinated. The chapter presents a theoretical overview of the three trade-offs and what their potential implications are.

Chapter 2 (**Tavasszy**) discusses the decarbonization of freight transport systems through internalizing the environmental costs of transport, by means of carbon emission-based taxes. The impact mechanisms of carbon taxes are explored by investigating the reorganization responses of freight decision-makers. A series of empirical cases of recent modelling studies at city, corridor, country, continent, and global levels are used to draw lessons concerning the impacts of logistics decisions on the policies and identify needs for further research which includes decisions in specific areas or logistics, but also on decision processes, to better understand the dynamics of impact pathways.

Chapter 3 (**Sharif Azadeh, Maknoon, Chen, Bierlaire**) discusses the impact of collaborative scheduling and routing for interconnected logistics system which plays an important role towards having a more sustainable green freight transport. Building on the concept of Physical Internet (PI), the chapter studies the last mile delivery as well as vehicle dispatching problems under the assumptions of collaborative supply chain networks. An efficient resource planning with the minimal number of empty vehicle movements running on roads that ultimately leads to decrease carbon dioxide emission is proposed and tested by real data successfully.

Chapter 4 (**Salimi**) discusses the challenges of implementing CE in agri-food systems. The chapter starts with a discussion on the need of using CE in agri-food systems. Then by identifying the characteristics of agri-food systems, the challenges that the actors in these systems face provide solutions from CE principles to handle those challenges. It further discusses some of the protentional policies, practices, and decisions for having circular agri-food systems. These include AgroCycle for a circular economy, optimal use of all biomass, local food system, and waste management. Following these strategies, practices and policies facilitates the implementation of CE in agri-food systems and the creation of new opportunities for value-added activities in agri-food systems by focusing mainly on food production, consumption, and waste management.

Chapter 5 (**Kheybari, Pooya**) discusses the location problem of bioethanol distribution centers based on two objectives: optimally locating bioethanol distribution centers and finding the minimum number of distribution centers. A hybrid methodology consisting of the best-worst method (BWM) and a set coving model is presented. The performance of the proposed methodology is evaluated by a set of data collected from Iran. The methodology proposed in this chapter can be used in several other similar contexts where geographical distribution of the alternatives and their interconnectedness needs to be incorporated in making the decision.

Chapter 6 (**Lami, Bottero, Abastante**) discusses the use of the multi-criteria decision analyses (MCDA) in urban and territorial decision processes. The reflections provided by this chapter come from a critical analysis of different case studies in urban and territorial planning realms, towards a sustainable development. The applications presented show that the MCDA can be a useful support to the decision-makers in order to structure the decision process in question, characterized by a plurality of stakeholders with different interests, powers, and goals. In particular, starting from the case studies, the authors highlight the applicability and the decision-making relevance of the different MCDA.

Chapter 7 (**Fallah Lajimi**) discusses the relationship between buyer and supplier in improving the sustainability of supply chain. The chapter introduces a process for the segmentation of the sustainable suppliers and presents strategies for collaboration and improvement of the suppliers. Sustainable suppliers are assigned into seven segments (three main segments, economic, social, and environmental; three balancing segments, bearable, viable, and equitable; and supplementary segment, sustainable). Finally, the improvement and development strategies appropriate to each dimension of sustainability are suggested.

Chapter 8 (**Gupta, Barua**) discusses internal barriers to green innovation through literature review. A hybrid of three different methodologies is employed. First, ISM (Interpretive Structural Modelling) is employed to identify the relationship between these barriers. ISM also helps in identifying driving barriers that have the most effect on the system. Next, using BWM (best-worst method), the driving barriers are ranked. In the third step, VIKOR (VlseKriterijumska Optimizacija I Kompromisno Resenje) is applied to rank the ability of manufacturing organizations in overcoming these barriers.

Chapter 9 (**van de Kaa**) discusses the determinants of information overload and its effect on decision quality. This chapter has a unique focus on some aspects of social sustainability which relate to the health of people working in an organization. The chapter draws from decision theory and the theory of human information processing and focuses on the individual characteristics of decision-makers and the quality of the information provided by the information system. Seven propositions are formulated and several remedies for information overload are discussed. A conceptual model is developed to provide a better understanding of the concept of information overload.

References

Geng, Y., Sarkis, J., & Ulgiati, S. (2016). Sustainability, well-being, and the circular economy in China and worldwide. *Science, 6278*(Supplement), 73–76.

Kahneman, D., & Tversky, A. (2013). Prospect theory: An analysis of decision under risk. In *Handbook of the fundamentals of financial decision making: Part I* (pp. 99–127). World Scientific.

Kates, R. W., Clark, W. C., Corell, R., Hall, J. M., Jaeger, C. C., Lowe, I., McCarthy, J. J., Schellnhuber, H. J., Bolin, B., & Dickson, N. M. (2001). Sustainability science. *Science, 292*(5517), 641–642.

Rezaei, J., Papakonstantinou, A., Tavasszy, L., Pesch, U., & Kana, A. (2019). Sustainable product-package design in a food supply chain: A multi-criteria life cycle approach. *Packaging Technology and Science, 32*(2), 85–101.

Rowe, A. J., Boulgarides, J. D., & McGrath, M. R. (1984). *Managerial decision making*. Citeseer.

Sauvé, S., Bernard, S., & Sloan, P. (2016). Environmental sciences, sustainable development and circular economy: Alternative concepts for trans-disciplinary research. *Environmental Development, 17*, 48–56.

Simon, H. A. (1979). Rational decision making in business organizations. *The American Economic Review, 69*(4), 493–513.

Slovic, P., Fischhoff, B., & Lichtenstein, S. (1977). Behavioral decision theory. *Annual Review of Psychology, 28*(1), 1–39.

Contents

About the Editor

Jafar Rezaei is Associate Professor and Head of section Transport and Logistics at the Faculty of Technology, Policy and Management, Delft University of Technology (TU Delft), The Netherlands. He obtained his PhD in Supply Chain Management from TU Delft in 2012 and has a background in Operations Research. He has published in several scientific journals in Operations Management and Operations Research mainly on the topics of decision-making in different areas including (sustainable) supply chain management and logistics. He serves as the Editor-in-Chief of the *Journal of Supply Chain Management Science* and as a member of the editorial board of several other scientific journals. His main research interests are decision-making methods and (sustainable) logistics and supply chain management. One of his outstanding achievements is the development of a decision-making method called Best-Worst Method (BWM), which has received significant attention by researchers and practitioners in many different scientific and application areas.

Chapter 1
Logistics in the Circular Economy: Challenges and Opportunities

Alistair Beames, G. D. H. Claassen, and Renzo Akkerman

Abstract Circular economy (CE) is a concept that has gained considerable attention in recent years, particularly in the domain of *Industrial Ecology*. CE requires products to be easily repaired, refurbished, remanufactured, and eventually recycled. The transition to a CE creates distinct material flows that have to be managed in an efficient and sustainable manner. Existing studies on CE tend to focus on product design, material use, and the market potential of CE products with little attention paid to the logistics challenges associated with such developments. From a logistics perspective, CE can be seen as the integrated management of forward and reverse flows of products in a supply chain. In the operations and supply chain management literature, a large body of knowledge on how to operationalize closed-loop supply chains (CLSCs) already exists and is a starting point for understanding logistics in the CE context. As with traditional forward supply chain network design, CLSC and CE supply chains also require decisions on the role of facilities, their location, their capacity allocation, and their demand and supply allocation. The CE concept does however introduce new challenges especially as circular business innovations converge to increased servitization and to more collaborative and open business models. The transition towards circular business models requires businesses to position themselves according to three key strategic decision-making problems, namely, (1) the extent to which the logistics network is centralized, (2) the extent to which the product is servitized, and (3) the extent to which logistics services are coordinated. This chapter presents a theoretical overview of the three trade-offs and what their potential implications are.

Keywords Logistics · Circular economy · Closed-loop supply chain · Product design · Supply chain network design

A. Beames (✉) · G. D. H. Claassen · R. Akkerman
Operations Research and Logistics, Wageningen University and Research, Wageningen, The Netherlands

J. Rezaei (ed.), *Strategic Decision Making for Sustainable Management of Industrial Networks*, Greening of Industry Networks Studies 8,
https://doi.org/10.1007/978-3-030-55385-2_1

1 Introduction

The circular economy (CE) is both a critique of what its advocates refer to as the 'linear economy' and a concept that promotes the preservation of the natural environment by minimizing resource extraction and waste generation. As opposed to disposing of products at the end of their use, CE seeks to extend the life of products and the materials embodied in products. Products that reach the end of their use-phase should then be re-used, remanufactured, and recycled as far as possible to prevent or postpone eventual incineration and landfilling. According to the Ellen MacArthur foundation (2015), re-use should be prioritized over remanufacturing which in turn should be prioritized over recycling. In other words, a hierarchy of product end-of-life processes exists, with the intention of preserving the embedded effort in the original product and the embedded energy in the material composition of the original product.

Circular business models recover value from used products and the materials embodied in products. This circular view on product manufacturing, product use, and end-of-life return options is illustrated in the Ellen MacArthur butterfly diagram (2015), shown in Fig. 1.1. As can be seen in Fig. 1.1, a distinction is made between the biological cycle (on the left) and the technical (on the right). The biological cycle includes products created from materials derived from living organisms but not fossil resources. The biological cycle is synonymous with the biobased economy.

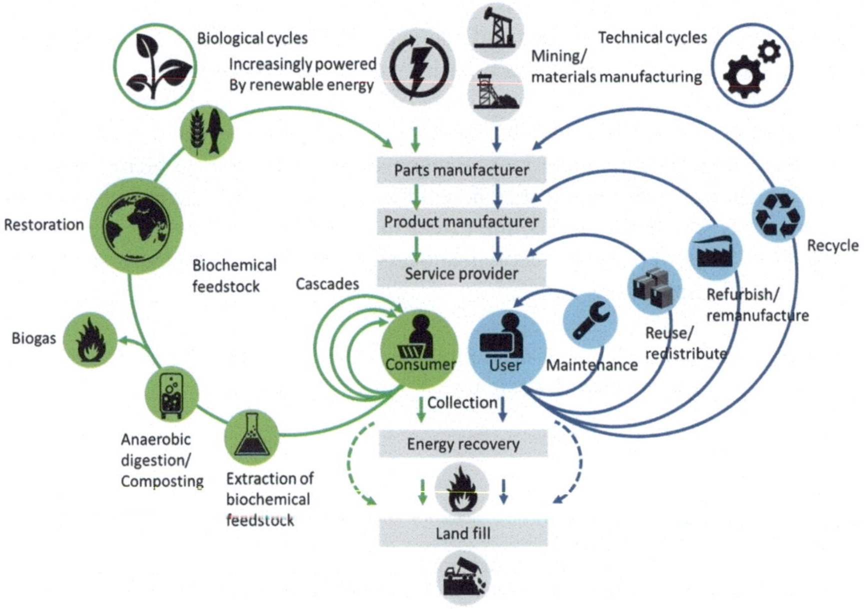

Fig. 1.1 Butterfly diagram illustrating a circular economy. (Adapted from the Ellen MacArthur Foundation 2015)

The technical cycle, however, includes products created from metals and minerals that initially need to be extracted from the subsurface. Ideally, circular products should all be derived from biological sources although this is not possible for many technologies.

Traditional linear product supply chains are often designed without taking the handling and disposal of products into account. The traditional product has a definite beginning, which is the sourcing of raw materials and the manufacture of the product. This can be referend to as the 'pre-use phase'. The customer uses the product, the 'use-phase', before the product is disposed of. The end of the product life can be referred to as the 'post-use phase'. In the CE, the end-of-life or post-use phase is the point at which the product or material is returned to the manufacturer so that it can be repaired, remanufactured, or recycled and then redistributed. In principle the sourcing of new materials for new products are displaced. It is important to point out, however, that according to the second law of thermodynamics, waste streams can be minimized but as long as the system is not isolated, there will always be a waste fraction that can no longer be recycled. This fraction will eventual need to be replaced by virgin materials. What can no longer be recycled can be sent energy recovery or used as construction material.

The role of designers and material engineers in the CE is to develop robust products that:

1. Extend the product use-phase for as long as possible to reduce the need for maintenance and the need for replacing products. In turn longer lasting products avoid all the associated environmental burdens (such as the depletion of limited resources and emissions) of collecting and processing used products and all the environmental burdens associated with manufacturing and distributing replacement products.
2. Maximize the potential for re-use, remanufacturing, and eventual recycling so that waste is essentially designed out of the system.

The role of logistics and supply chain management is to facilitate the efficient flow of materials in a circular economy. In other words, logistics and supply chain management provides the framework upon which every step in the life of the product either side of the use-phase can be optimized. The framework enables strategic decision-making with regard to:

(a) Material sourcing decisions, collaborating with manufactures of other products and utilizing their waste streams as inputs
(b) Typical facility location-allocation problems as well as the location-allocation of facilities for reverse logistics, maintenance, reprocessing, and redistribution
(c) Routing of distribution, collection, and redistribution networks
(d) Determining the scale at which product revenues exceed costs

The evolution of circular product design, new product life cycle management approaches, and various product recovery options has led to many structural changes in logistics processes. This chapter provides a logistics and supply chain perspective on the transition to a CE. The following sections explain the interconnectivity of circular product design and supply chain network design with reference to three key logistics and supply chain strategic decision-making problems relevant for the design and operation of circular businesses.

2 Logistics in the Circular Economy

A supply chain consists of all parties involved, directly or indirectly, in the flow and transformation of goods and services from the origin of the product through to the customer. Logistics involves the business activities required to match the consumer demand for products from businesses with the supply of products. These activities span throughout product supply chains, from raw material extraction through to final product delivery. The collection and processing of used products and waste also falls within the definition of logistics. Logistics does not only involve transport but also relates to (1) strategically designing a supply chain in terms of the scale and location decisions, (2) the matching of supply and demand with the resulting networks, and ultimately (3) the detailed planning of the flow of products through the supply chains. In other words, logistics deals with getting the right product at the right place, at the right time, in the right amounts, and of the right quality.

The hierarchy of end-of-life processes (repair, re-use, refurbish, remanufacture, recycle, waste-to-energy, and landfilling) can be understood from these perspectives. Firstly, the potential for a product to be re-used, remanufactured, or recycled falls within the domain of circular product design, and it is essential that circularity is already considered in this early stage. Secondly, for products that have a significant use-phase, there is emphasis on optimal use of the product during its life cycle and possible extension of this life cycle. Maintenance or repair activities can reduce overall material requirements. Thirdly, all materials in products, whether designed for prolonged use, remanufacture, or recycling, will eventually be disposed of. The re-use, remanufacture, and recycling of products simply acts as a buffer, delaying the point in time when the materials will be unusable. The disposal of materials has traditionally been the responsibility of local governments, who either disposed of waste themselves or outsource waste collection and disposal activities to waste management companies. The end-of-pipe solutions used for disposal include dumping, sanitary landfill, and incineration. Energy in the form of biogas can be harvested from sanitary landfills. The preferred alternative to landfilling is incineration that recuperates energy, also referred to as waste-to-energy. As society shifts towards circularity, especially with longer-lasting products that can be re-used, the quantities of waste managed by municipalities and waste management companies will decrease. Finally, logistics determines how the materials required for the product are brought together, how the product is brought to the consumer, and how used products are either brought back to the manufacturer or disposed of. These perspectives on the circular economy are also represented by Achterberg et al. (2016) in the Value Hill, which illustrates the transition from a linear economy to a circular economy. The Value Hill includes life cycle extension in the use phase, the many

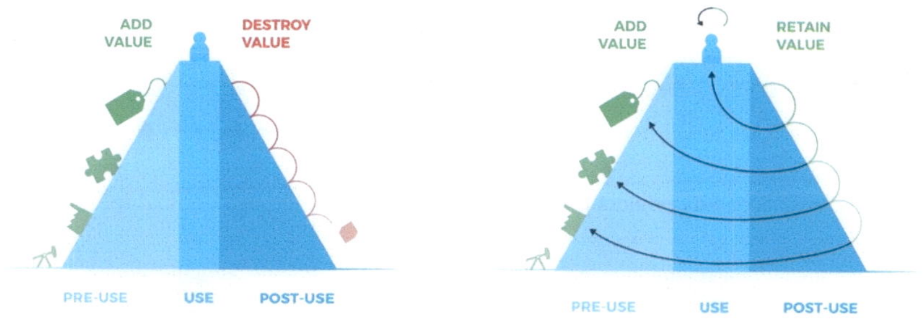

Fig. 1.2 Value Hill diagram by Achterberg et al. 2016. The pyramid on the left represents the linear economy where value is destroyed post use. The pyramid on the right represents the circular economy where value is retained post use. The greatest value retention occurs at the pinnacle where only maintenance is required

reverse flows in the post-use phase, as well as an emphasis on organizing logistics and supply chain activities across the different phases (Fig. 1.2).

3 Circular Product Design and Supply Chain Network Design

The basis of designing circular supply chains is similar to that of conventional supply chains. The basis of supply chain network design requires making decisions within each of the following four categories (Chopra and Meindl 2016): facility role, facility location, capacity allocation, and demand and supply allocation. The facility is where products are produced or where materials are stored or transferred. Each type of facility in the chain needs to be determined. For circular products, this may also include collection hubs, refurbishment facilities, and disassembly facilities. The locations of each type of facility need to be determined. Transporting materials and eventual products is an expense, and whether the product is circular or not, the distance between facilities should be as short as possible. There is, however, a trade-off that needs to be made between short network distances and the extent to which facilities can be decentralized. For each facility and location, the capacity needs to be defined. The capacity of each facility is also a function of how many facilities are in the supply chain network.

Designing the supply chain also requires determining the share of each market serviced by each facility. Retailer chains, for example, need to understand the size of the catchment area that each of their stores are servicing. The catchment size in turn, also determines the capacity at which the nodes upstream should be operated at. This also concerns the supply side, especially when production is based on reusable or recyclable material. On the supply side, the catchment area represents the material supply base.

Conventional supply chains can be referred to as open-loop supply chains and involve the flows of material and eventual products from the producer to the

Fig. 1.3 Linear economy product life cycle from manufacture, through use to disposal

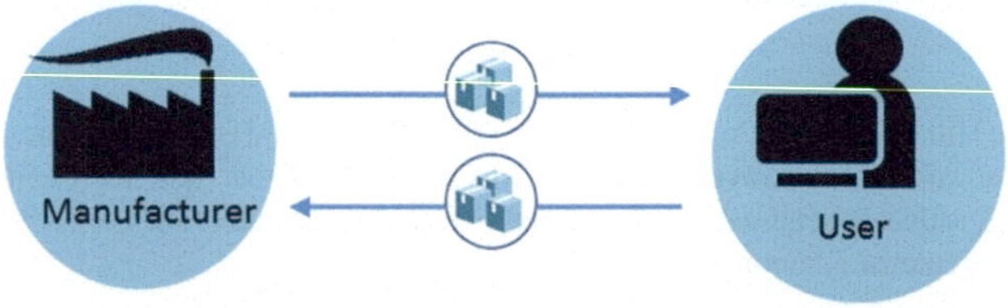

Fig. 1.4 Circular network design principles to reduce resource use (adapted from Bocken et al. 2016). Circular products last longer and therefore need to be replaced less often. Circular products are made from materials, which have less of a burden on the environment. Finally, circular products are returned to the manufacturer as opposed to being disposed of

customer. The supply chain is open because the product leaves the initial supply chain once it reaches the customer. At the end of the product's use-phase, it is then collected and either re-used, remanufactured, recycled, or disposed of via a different supply chain to the supply chain which delivered the product in the first place. If the supply chain also includes the collection of the used products, then it is referred to as a closed-loop supply chain.

From a logistics perspective, CE can be seen as the integrated management of forward and reverse flows of products in a supply chain. In the quantitative operations and supply chain management literature, a large body of knowledge on how to operationalize closed-loop supply chains (CLSCs) already exists. CE literature, mostly from *Industrial Ecology*-based journals, prescribes sets of principles predominantly focused on the design and material composition of products. Less emphasis is placed on supply chain network design. Bocken et al. (2016) do however describe three design principles that could underpin the circular products supply chain network design, namely, (1) slowing, (2) narrowing, and (3) closing (material) loops. Figures 1.3 and 1.4 illustrate the three principles. Figure 1.3 is of the traditional linear economy product, which is manufactured, delivered to the consumer and disposed of when the product reaches its end-of-life. The product is also replaced multiple times. From a logistics perspective, this product supply chain can be referred to as open loop. Figure 1.4, depicts a circular product supply chain or in logistics terms, a closed loop supply chain. The circular product lasts longer and therefore does not need to be replaced. The transport of the product has a smaller impact on the environment because it is constructed out of lighter (as well as more environmentally friendly) materials. Finally, the circular product is returned to the manufacture at its end-of-life, for refurbishing, remanufacturing, or recycling.

Slowing material flows involves product life cycle extension. In other words, designers should design products that are durable and last longer and allow for the possibility of servicing, repair, and remanufacture. Longer-lasting products implies that the need for replacing products is reduced. From a logistics perspective, on the one hand, this means less deliveries and less transport of new products. On the other hand, the need to transport new products would be replaced by the reverse flows of products requiring maintenance, repair, and remanufacturing. Waste collecting and processing will also be reduced by extending the life of products (including packaging). In many cases, product ownership will also change; it will be retained by product manufacturers who simply lease their products to their customers. In other words, 'product-as-a-service' will replace customer product ownership. This development is called servitization and incentivizes longer-lasting products and the efficient maintenance of products on the side of the product owner (the manufacturer).

Narrowing material flows also starts with the product designer, designing products in such a way as to minimize material inputs. This could include fewer materials used in the manufacturing of the product and less energy required in the operation of the product after manufacture. In other words, the products are environmentally less damaging or "lighter." Part of this discussion is the possible minimization of transport distances of the resources chosen by the product designer. CE principles generally also advocate reducing dependencies on imports (Geissdoerfer et al. 2018), since such dependencies present resource security risks. Stahel (2013) also argues that the repair, remanufacture, and recycling flows should be kept small and local in order to avoid products being transported back and forth over long distances.

Closing material flows begins with a switch from landfilling and incinerating products and residual waste to the separation of waste streams and recycling. Ultimately, recycling should be replaced by reverse logistics, and every product manufacturer will therefore have to have a closed-loop product system. From a supply chain perspective, reverse logistics has already been studied for many years (e.g., Govindan et al. 2015; Govindan and Soleimani 2017), and many opportunities to apply and extend this knowledge base exist. Closing loops can also be extended across different supply chains, where products or waste streams are not necessarily re-used in their own supply chain, but by another stakeholder that can make use of the material. An example of such re-use is industrial symbiosis, which is the use of another company's waste flows as a feedstock (see also Herczeg et al. 2018).

In the current transition towards the CE, many circular business models are being initiated by start-ups and by existing businesses that traditionally provided waste management services. Initiatives range from growing mushrooms on collected used coffee grounds to the production of biogas from collected municipal organic waste. Circular companies that are able to valorize waste or side streams need to be able to compete in the market with producers that use conventional feedstock and fossil fuels. Whether or not these circular businesses can compete in terms of price per unit is to a large extent dependent on whether logistics and production costs can be kept low enough to achieve a competitive market price for the product being made. Three strategic decision-making problems specific to circular businesses need be considered in achieving an optimal supply chain.

4 Key Logistics Strategic Decision-Making Problems in a Circular Economy

The optimal supply chain network configuration for circular products requires companies to consider their positioning on the following three strategic decision-making problems:

1. Centralized versus decentralized production and logistics
2. Product-oriented versus service-oriented sales
3. Coordinated versus collaborative organization

Each of the three strategic decision-making problems is explained in the following subsections.

4.1 Centralized Versus Decentralized Production and Logistics

The centralized production or processing of materials, whether that be for forward logistics (supply new products to customers) or reverse logistics (collecting used products), is generally more efficient than smaller decentralized facilities. Depending on the type of processing technology required, a larger facility that handles more product has lower start-up or fixed costs than multiple facilities handing the same quantity of product. The efficiency achieved with larger facilities can be described as *economies of scale*. Larger facilities with larger throughput allow for gains in economic efficiency by decreasing average total costs per unit produced in the long run. The gains in efficiency are due to greater labor and managerial specialization as well as allowing assets like machinery to be utilized continually, as opposed to intermittently or lying idle. Large-scale production also allows for bulk purchasing of inputs, which also reduces average total costs per unit.

The greater efficiency of centralized processing and production, as opposed to smaller-scale decentralized production, comes at the cost of a more sparsely distributed network and therefore greater transport costs. For example, having multiple warehouses in a decentralized network, as opposed to a single warehouse in a centralized network, from which to distribute products to retail outlets in a city, reduces the transport distances between the warehouses and the retail outlets. The same principle applies to processing facility for the reserve flow of products. A single biogas plant, invariably, requires greater transport distance than a more distributed or decentralized network of smaller digesters. Designing the optimal network requires finding a balance between the fixed costs of each facility and the variable costs of transport to and from each facility.

Other important considerations, which are particularly relevant to end-of-life processes (re-use, remanufacture, recycle, waste-to-energy, and landfilling), are the quantities, values, and geographic dispersion of the used products or waste streams. The more value retained in the used product, the greater the distance over which it can be transported, and the smaller and more geographically dispersed the products can be. For example, used precious metals retain their value, and therefore the cost of transport is relatively lower than, for example, manure, which has very little value. In other words, the value of the used product should cover the transport costs, and if this margin is very small, then a more decentralized network might be necessary.

Chen et al. (2012) evaluated 88 recycling projects in 23 towns in Japan to draw conclusions about the optimal scale of production for different value streams. They concluded that materials, such as oil, metals, plastics, paper, and Waste Electrical and Electronic Equipment (WEEE), could be handled across a dispersed regional network of recycling and reprocessing facilities, since the market value of the materials cover the transport costs. Organic waste, mixed municipal solid waste (MSW), and demolition waste are however of relatively little value per unit of mass, and therefore only a more local decentralized network of recycling and processing can allow for the transport costs to be covered.

Finally, product quality concerns might also impact the level of centralization; in the case of biological materials, there might be limited time in which the disposed products have to be processed, or its value might be lower (and, for instance, only be suitable for a waste-to-energy treatment).

The opportunity for logistics sector lies in designing networks that strike a balance on the continuum between centralized production with economies of scale on the one hand and decentralized production and transport cost minimization on the other hand.

4.2 Product-Oriented Versus Service-Oriented Sales

The circular economy literature discusses a change in traditional business operating models, from the business-to-customer transfer of product ownership to firms keeping ownership of products and instead providing the service that the product would otherwise deliver. The transition typically involves a switch from delivering value in the form of a finished product to delivering value in the form of performance-based services (Reim et al. 2015). As opposed to the customer buying a product and owning it until its end-of-life, the firm retains ownership of the asset and receives revenue for the value delivered by the asset to the customer. The firm, is therefore, incentivized to operate and maintain the asset as efficiently as possible. The transition goes beyond Extended Producer Responsibility (EPR) schemes, in which the producer is merely responsible for product's end-of-life but instead businesses define their business model according to the extended life of assets

(Van Engeland et al. 2018). The new type of business models is defined as ‘product-as-a-service’, ‘product-service’, or product service systems (Reim et al. 2015; Annarelli et al. 2016; Franco 2017; Geisendorf and Pietrulla 2018)

Businesses are also therefore required to balance the trade-off between extending the life of existing assets and replacing existing assets with new more efficient technology. A classic example of the product-as-a-service operating model is the leasing of photocopy machines, where the leasing company is responsible for maintenance and the continuous operation of the photocopy machines. Instead of the customer buying a photocopy machine, they are ‘paying per use’. The business model can be referred to as ‘use-oriented’. As with all assets, over time, it might be more economical to replace existing machines with newer technology. The firm is incentivized to carefully weigh up the cost and benefits of maintaining existing assets against new investments, and in this sense, the onus of the material efficiency of products or assets remains with the initial manufacturer. The question is how incumbent firms ought to make this transition and where the relevant opportunities are for the logistics sector.

Frishammar and Parida (2019) define a roadmap for firms to transform linear business models to circular business models. The roadmap is based on empirical observations of how eight existing firms, motivated by consumer interest in sustainable business practices on the side of firms, have made this transition. Frishammar and Parida (2019) were able to deduce a four-phase approach to transforming from typical linear business models to greater circularity. Phase 1 involves identifying the potential for adopting circular business principles and creating awareness regarding circularity amongst the existing customer base. If opportunities can be identified, then the phase 2 can be initiated in which the existing business model is exhaustively evaluated in terms of shortcomings, barriers, and potential circular business opportunities. This phase also involves creating awareness within the firm regarding the potential for circularity. The shortcomings and opportunities are the starting point for phase 3, in which the new circular business model is designed. Here, incumbent firms can draw from examples of case studies of other firms who have already made the transition. The firm in question needs to achieve internal alignment so that all the necessary departments in the firm are aligned in supporting the transition. The firm also needs to identify potential for collaboration with other firms or “ecosystem partners” and other network actors. Finally, the phase 4 involves validating the proposed transformation by introducing a pilot-scale prototype. The small-scale trial allows the firm to identify room for improvement and potential pitfalls of the circular business design before moving towards large-scale rollout.

The logistics sectors in many cases are ecosystem actors that provide the operational infrastructure for moving products between producer and customer. As firms continue towards increased servitization, the one-time flows of products to customer and from customers to end-of-life treatment (disposal, recycling etc.) will be replaced with products (or assets) remaining at customers for longer and servicing and maintenance of those assets. In other words, there will be less throughput of product and this will be replaced by the movement of service personnel and spare parts to and from customers. Logistics service providers have the opportunity to more efficiently manage maintenance scheduling. A further evolution towards circular economy would involve business-to-business servitization of assets, where firm share centralized production and manufacturing infrastructures. In this case, logistics services providers also stand to gain in terms of greater business-to-business flows between business locations.

4.3 Coordinated Versus Collaborative Organization

Collaborative logistics is the sharing of supply chain and distribution network infrastructures between multiple companies to reduce logistics costs. Collaborative logistics can be vertical or horizontal. Vertical collaboration involves different companies managing different consecutive stages along the same supply chain. Horizontal collaboration involves different companies that are usually competitors sharing each other's logistics infrastructure to reduce costs. Collaborative logistics can also include handing over logistics operations to an independent party that does not share sensitive operational information between the parties involved but ensures that the part of the supply chain it is responsible for operates optimally. Allowing such a fourth party to manage operations can be referred to as coordination and leads to developments like supply chain control towers. Beyond reducing logistics costs, companies who can share an existing network also improve the environmental performance of their products (e.g. reducing travel distances and truck utilizations and therefore the associated air emissions of the product).

Circular products can be placed on a spectrum between various different firms working together, either in sequence or within the same stage of the supply chain and a fully coordinated supply chain. In the CE context, traditional product manufacturing can consider shifting from an open-loop supply chain, to closed-loop supply chain, in which case the product is returned from the customer to the initial producer, when the consumer no longer needs the product or when the product reaches the end of its life. The expansion of the supply chain to include a take back program of the product can be described as integration, which can also be done vertically or horizontally. Integration of downstream supply chain activities, such as shifting from an open loop supply chain to a closed loop supply chain would be forward integration. Horizontal integration between a firm and other firms operating in the same supply chain stage that would otherwise be competitors is also possible in closed-loop context, especially in situations where the management of reverse flows is regulated on industry level. A well-known example is the collection of Waste Electrical and Electronic Equipment (WEEE). Integration is an important concept because it describes the transition of firms from open-loop supply chain towards closed loop supply chains; however, integration is not always synonymous with greater collaboration.

Kortmann and Piller (2016) describe two ways in which businesses are adapting to consumer and stakeholder demands for more sustainable, as well as open, business practices. By open business practices, they refer to firms being transparent

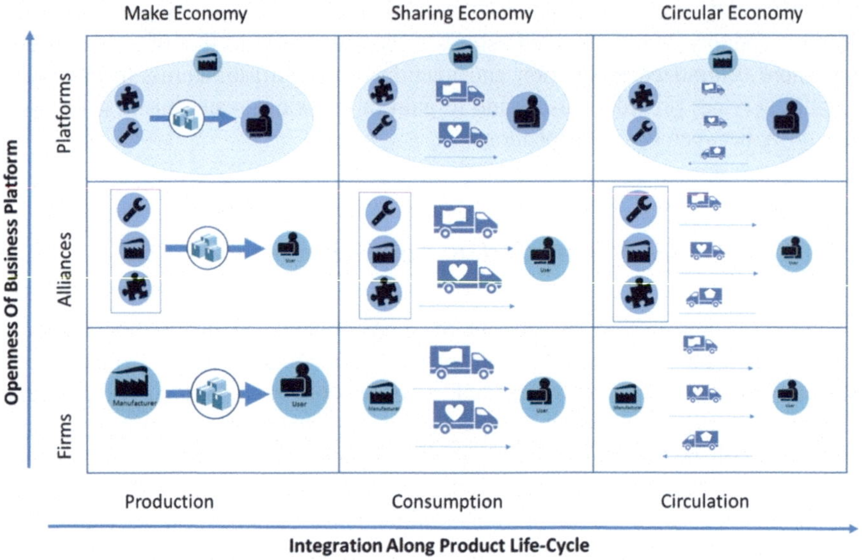

Fig. 1.5 Business model archetypes in the transition towards open and circular value chains. (Adapted from Kortmann and Piller 2016)

about business decisions and allowing stakeholders and customers to play a role in defining the business model. The shift towards greater sustainability lies in firms expanding the scope of the business operations to include taking back products from consumers (Fig. 1.5).

They map out nine archetypes of business models along two axes. The first or horizontal axis is the product life cycle continuum from the conventional forward value chain of linear business models towards the closed loop value chain of circular business models. The continuum is further divided into production, consumption, and circulation and in other words represents greater forward integration. The second or vertical axis is the business openness continuum from the conventional firm operating in isolation towards full collaboration and in other words greater horizontal integration. The continuum is divided into firms, alliances, and platforms.

From a logistics perspective, Archetype 1 is the traditional single actor, open-loop supply chain. Archetype 9, on the other side of both continuums, a circulation-platform operator, is synonymous with a fourth party responsible with cross-chain coordination. Kortmann and Piller (2016) show that firms need to cater to consumer demands for greater sustainability in terms or circulation and greater openness and this will also present opportunities for logistics sector.

5 Conclusions

This chapter provided an introduction to the circular economy concept and specifically what the concept entails regarding product design, product life cycle extension, as well as a hierarchy of end-of-life possibilities (re-use, remanufacturing, and recycling). From a logistics perspective, circular economy is predominantly about a shift that has been conventionally referred in logistics literature as open-loop supply chains, towards closed loop supply chains. In other words, the traditional life cycles of products from manufacturing, product use, to product disposal is being replaced with manufacturing, product use, and take back. In this sense, there is both a forward and reverse flow of products between producers and consumers. This also means less products will be treated as waste by local government, and instead manufacturers will displace landfill and incineration with longer lasting products, re-use, and remanufacturing schemes. The logistics sector will play a critical role in orchestrating the newly established flows between the different ecosystem actors.

The challenges that arise include:

1. Creating affordable reverse take-back programs and reverse logistics that support geographically dispersed materials and products
2. Reducing idle products and assets by firms that invest in product-as-a-service business models but also bear the risk of those products and assets becoming outdated
3. Sharing logistics networks with competitors at the risk of losing market share to competitors

The opportunities that arise including:

1. Sourcing cheaper material inputs locally that would otherwise be waste
2. Business-to-business servitization of assets reduces capital expenditure and therefore barriers to new markets for entrepreneurs
3. Circulation-platforms and open business models allow users to be directly inform product research and development

The perspective on how supply chain networks ought to be designed and operated differs between the strictly product design-oriented literature from the Industrial Ecology domain and the well-established knowledge of closed loop supply chains in the logistics domain. From a product design perspective (e.g., Stahel 2013; Bocken et al. 2016), flows should be slowed, narrowed, and closed. In other words, the flow of products should be reduced by designing products to last longer, using less environmentally intensive materials, and ensuring that products are returned to the producer at their end-of-life. From a logistics and (closed-loop) supply chain perspective, business models in the circular economy should consider three key strategic decision-making problems in the design and operation of their supply chain: (1) the extent to which the network is centralized or decentralized; (2) the extent to which a product is servitized; and (3) the extent to which the supply chain activities requires a collaborative effort. A perfect circular business model and corresponding production and logistics network does not exist, as many case-specific aspects need to be considered. The three strategic decision-making problems outlined in this chapter provide a framework for achieving greater circularity and supply chain integration.

Acknowledgments This chapter is based on the work performed in the LogiCE project, funded by NWO, the Dutch Research Council, under grant number 439.16.611. The authors thank their project partners for their collaboration and for their contributions to interesting discussions on the role of logistics in a circular economy.

References

Achterberg, E., Hinfelaar, J., & Bocken, N. (2016). *Master circular business with the value hill* (Research Report). Amsterdam: Circle Economy & Sustainable Finance Lab.

Annarelli, A., Battistella, C., & Nonino, F. (2016). Product service system: A conceptual framework from a systematic review. *Journal of Cleaner Production, 139*, 1011–1032.

Bocken, N. M., de Pauw, I., Bakker, C., & van der Grinten, B. (2016). Product design and business model strategies for a circular economy. *Journal of Industrial and Production Engineering, 33* (5), 308–320.

Chopra, S., & Meindl, P. (2016). *Supply Chain Management. Strategy, Planning & Operation.* Boston: Pearson Prentice Hall.

Chen, X., Fujita, T., Ohnishi, S., Fujii, M., & Geng, Y. (2012). The impact of scale, recycling boundary, and type of waste on symbiosis and recycling: An empirical study of Japanese eco-towns. *Journal of Industrial Ecology, 16*(1), 129–141.

Franco, M. A. (2017). Circular economy at the micro level: A dynamic view of incumbents' struggles and challenges in the textile industry. *Journal of Cleaner Production, 168*, 833–845.

Frishammar, J., & Parida, V. (2019). Circular business model transformation: A roadmap for incumbent firms. *California Management Review, 61*(2), 5–29.

Geisendorf, S., & Pietrulla, F. (2018). The circular economy and circular economic concepts—a literature analysis and redefinition. *Thunderbird International Business Review, 60*(5), 771–782.

Geissdoerfer, M., Morioka, S. N., de Carvalho, M. M., & Evans, S. (2018). Business models and supply chains for the circular economy. *Journal of Cleaner Production, 190*, 712–721.

Govindan, K., & Soleimani, H. (2017). A review of reverse logistics and closed-loop supply chains: a Journal of Cleaner Production focus. *Journal of Cleaner Production, 142*, 371–384.

Govindan, K., Soleimani, H., & Kannan, D. (2015). Reverse logistics and closed-loop supply chain: A comprehensive review to explore the future. *European Journal of Operational Research, 240* (3), 603–626.

Herczeg, G., Akkerman, R., & Hauschild, M. Z. (2018). Supply chain collaboration in industrial symbiosis networks. *Journal of Cleaner Production, 171*, 1058–1067.

MacArthur, E., Zumwinkel, K., & Stuchtey, M. (2015). *Growth within: A circular economy vision for a competitive Europe*. London: Ellen MacArthur Foundation.

Kortmann, S., & Piller, F. (2016). Open business models and closed-loop value chains: Redefining the firm-consumer relationship. *California Management Review, 58*(3), 88–108.

Reim, W., Parida, V., & Örtqvist, D. (2015). Product–Service Systems (PSS) business models and tactics – A systematic literature review. *Journal of Cleaner Production, 97*, 61–75.

Stahel, W. R. (2013). Policy for material efficiency—sustainable taxation as a departure from the throwaway society. *Philosophical Transactions of the Royal Society A, 371(1986)*, 20110567.

Van Engeland, J., Beliën, J., De Boeck, L., & De Jaeger, S. (2018). Literature review: Strategic network optimization models in waste reverse supply chains. *Omega*. in press.

Chapter 2
The Influence of Logistics Decisions on Transport Decarbonization: Lessons from Local to Global Scale

Lóránt Tavasszy

Abstract One of the proposed approaches to decarbonize freight transport systems is to internalize the environmental costs of transport, by means of carbon emission-based taxes. The expected impact is a reduction of transport demand and an increased use of environmentally friendly transport technologies. The magnitude of the impact will depend on the levels of taxation, the activities to which they apply and the degrees of freedom allowed to companies to react to the taxes. In this chapter we explore the impact mechanisms of carbon taxes by investigating the reorganization responses of freight decision-makers. We do this through a series of empirical cases of recent modelling studies at city, corridor, country, continent and global level. The city case involves an application of an agent-based model to evaluate a carbon credit point system for city logistics. The corridor case involves carbon pricing of container transport in the hinterland of the port of Rotterdam over a multimodal network. The country and continental cases describe the effects of network-wide truck charging, with a focus on mode choice, vehicle type and routing. The global case concerns a full economic impact analysis of internalization of external costs of supply chains, also looking at the effect of changes in sourcing decisions of companies. We draw lessons concerning the impacts of logistics decisions on the impact of policies and identify needs for further research. Common findings relevant for climate change policy include the following: (1) prices needed to achieve a significant impact are a multiple of current market prices, (2) logistics decisions may act as buffer for the propagation of taxes towards consumers and, as a result, (3) the ultimate price impacts for consumers could remain small. In order to be able to predict impacts of climate policies, there is a need to continue research on the way companies take logistics decisions. This includes decisions in specific areas or logistics, but also on decision processes, to better understand the dynamics of impact pathways.

L. Tavasszy (✉)
Delft University of Technology, Delft, The Netherlands
e-mail: L.A.Tavasszy@tudelft.nl

J. Rezaei (ed.), *Strategic Decision Making for Sustainable Management of Industrial Networks*, Greening of Industry Networks Studies 8,
https://doi.org/10.1007/978-3-030-55385-2_2

Keywords Logistics · Transport · Decarbonization · Transport policy · City logistics · Container transport · Global supply chain

1 Introduction

The challenge of decarbonizing the global logistics system is a formidable one. The targets for decarbonization were framed by 195 countries in Paris during the 2015 COP21 meeting. Limits have been set on allowed global warming temperatures, and countries have subsequently translated this into national agreements to reduce their carbon emissions. In Europe, the intention was stated to achieve a 60% reduction of carbon emissions by 2050, compared to the 1990 levels. Given that emissions have only increased since then, and still are, the current reduction target is more strict and stands at 83%. Figure 2.1 shows the development of this emission gap at global level, through the decades.

Our focus in this chapter is on the logistics system as contributor to the overall target. Various studies are available in the literature that explain the ways in which carbon reduction can be achieved. The five main strategic routes for decarbonization are as follows (we refer the interested reader to McKinnon (2018) for a detailed discussion):

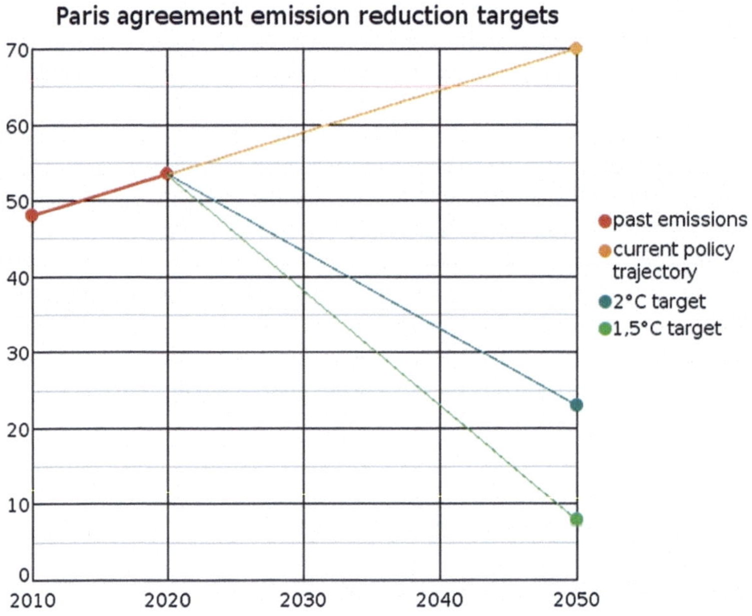

Fig. 2.1 Carbon emission gap. (UNEP 2016)

1. Reduce demand for transport (in volume, weight, distance).
2. Increase the use of least polluting transport modes.
3. Optimize the utilization of transport capacity.
4. Improve fleet energy efficiency.
5. Reduce the carbon content of energy used for transport.

To see that these routes are collectively exhaustive and mutually exclusive, it is helpful to understand the dimensions of each and how they are interlinked. The below equation, akin to the Kaya identity from environmental sciences (Kaya and Yokoburi 1997), demonstrates this point.

$$F = T \cdot M_{\mathrm{m}} \cdot U_{\mathrm{m}} \cdot E_{\mathrm{m}} \cdot C_{\mathrm{m}} \tag{2.1}$$

where

F = carbon emission [tCO_2][1]
T = demand for transport [tonne-km]
M_{m} = share of mode m
U_{m} = utilization of mode m [vehicles/tonne]
E_{m} = energy use of mode m [J/vehicle-km]
C_{m} = carbon intensity of energy source of mode m [tCO_2/J]

The design and implementation of measures to achieve the above changes is not a trivial issue. The logistics system is complex, with many interdependent stakeholders and much uncertainty about future demand and supply of service markets. Also, measures cannot just be forced upon the logistics sector. Private firms act autonomously and decide freely about the choices listed above. Up to a point, the objectives of the market will coincide with the decarbonization objective. Efficiency increases will be welcomed by the sector and will benefit the environment. As long as the measures do not put the demand for products and services at risk, and help to reduce costs, the sector will have a strong incentive to innovate. The costs of abatement of emissions with more radical innovations will be prohibitively large, however, with energy prices being too low to warrant investments in alternatives. Decarbonization beyond simple efficiency improvements will imply a cost increase and a possible reduction of competitiveness of a company. In order to reach the high decarbonization targets, additional measures will be needed to nudge markets to reduce carbon emissions further. One of these measures is the internalization of external costs of logistics, through carbon pricing.

The main policy question that we treat concerns the expected effects of carbon pricing policies on the options above. Our general expectation is that carbon pricing will stimulate change in all the above strategic directions. It could reduce the demand for carbon-intensive products and services and also allow more environment-friendly technologies to become competitive, driving out conventional technologies.

[1] CO_2 measured in metric tonnes abbreviated here as tCO_2 for the reader's convenience.

There are several approaches to implement carbon pricing through public policies, mostly using the instrument of government taxes. In addition, emission quota or so-called caps may be applied to determine the required level of the tax. If these quotas are company specific, their enforcement can be arranged via a system of permits or certificates, to allow a certain amount of carbon to be emitted. Eventually, permits can be made tradeable, to allow a re-distribution of emissions. In what follows, we will not go into these instruments in more detail and instead represent measures in a stylized way through price increases of logistics services.

Typically, such a policy question can be answered in some more detail using empirical, quantitative, predictive models of logistics systems, reproducing the decision-making of companies. Models can be built that include also dynamic behaviour of firms, so that one can evaluate whether the expected time to impact of carbon taxation is in line with the long-term targets. Modelling of behaviour of firms is not a trivial task. There is little generalized knowledge about how logistics decisions typically will respond to policies. To model them explicitly, we must take recourse to aggregate and stylized models of the freight system, econometrically estimated against observed decisions and flows (Tavasszy et al. 2019). We are not aware of any research about expected time duration of responses of firms to taxation measures; studies are generally static-comparative in nature.

Our objective is to contribute with a comparative discussion of a series of recent modelling studies at city, corridor, country, continent and global level. All the cases relate to the reorganization responses of logistics decision-makers, in the context of carbon pricing. The city case involves an application of an agent-based model to evaluate a carbon credit point system for city logistics. Decisions involved include order size, in-shop inventories, routing and truck type. The corridor case involves carbon pricing of container transport in the hinterland of the port of Rotterdam over a multimodal network. The country and continental cases describe the effects of network wide truck charging, with a focus on mode choice, vehicle type and routing. The global case concerns a full economic impact analysis of internalization of external costs of supply chains, also looking at the effect of changes in sourcing decisions of companies.

The paper is built up as follows. Section 2 introduces a conceptual framework of logistics reorganization measures that firms may take in response to decarbonization policies. We build on this framework of decisions throughout the paper, with empirical cases. In section 3, we present an overview of the cases, in terms of their main characteristics and the general expectations from the literature, concerning lessons to be learned. Sections 4, 5, 6, and 7 each describe a case in itself, at the scale of a municipality, a corridor, nations and the world, respectively. Section 8 synthesizes and discusses the findings across case studies, reaching back to expected lessons stated in Sect. 3. Section 9 concludes the paper with a summary of findings and some recommendations for research and policy.

2 Logistics Reorganization Responses to Carbon Pricing

The term "logistics reorganization response" was, to our knowledge, first suggested in the US Federal Highways Administration's (FHWA's) Freight Benefit-Cost Analysis Study (ICF & HLB, 2002). It concisely expresses the phenomenon that companies will reorganize to try to mitigate adverse effects or reap potential gains of policies on their business. This reorganization takes place by reviewing the relevant logistics decisions and implementing changes. Although the context of the FHWA study was to assess impacts following the principles of benefit-cost analysis, the systematic introduction it provides to impacts on firm decision-making is valuable for our cause. In their classification of effects of policies on society (Table 2.1), it is clear that the responses of firms are an important link in the impact chain of policies, between the primary, or first-order impacts, through business reorganization effects, until the ultimate external impacts that climate policies aim to reduce. In this paper, we are primarily concerned with the logistics decisions of firms and consumers that influence the second- and third-order benefits.

Our aim is to understand how climate policies propagate through the logistics system, by making reorganization responses explicit and studying their impact. The above classification notes the policies as cost-reducing transport policies, and the resulting effects as benefits, in line with the conventional aims of transport policies. In the case of climate change policies, however, the aim is to reduce external effects of transport, mostly by increasing prices or by enforcing the use of more expensive technology. Companies will deploy the same, second- and third-order reorganization measures as before, but will now direct them in a defensive way, to mitigate the impact of cost and price increases. If a company makes smart reorganization decisions, it will (1) reduce the impact of the price increase on the logistics costs of the firm, (2) make the company less sensitive to new increases and (3) reduce the necessary price increase of the service for its clients. Obviously, for policy makers, to be able to assess the final impacts of their climate policies, it is vital that these mechanisms are understood.

A simple example of a reorganization decision that illustrates well the above impacts concerns the choice of shipment size. If a receiver orders small shipments,

Table 2.1 Classification of societal benefits of transport policies (ICF & HLB, 2002)

First-order benefits	Immediate cost reductions to carriers and shippers, including gains to shippers from reduced transit times and increased reliability
Second-order benefits	Reorganization effect gains from improvements in logistics. Quantity of firms' outputs changes; quality of output does not change
Third-order benefits	Gains from additional reorganization effects such as improved products, new products or some other change
Other effects	Effects that are not considered as benefits according to the strict rules of benefit-cost analysis, but may still be of considerable interest to policy-makers. These could include, among other things, increases in regional employment or increases in rate of growth of regional income

Shaded area: focus of this paper

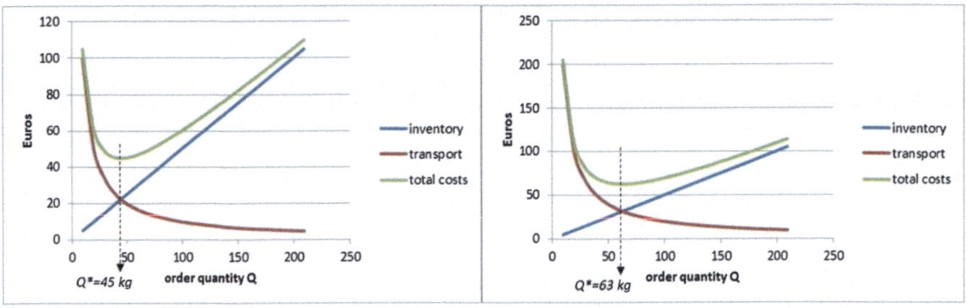

Fig. 2.2 Increase of transport price leading to a change in optimal shipment size Q*

he will typically incur relatively high transportation costs. Not only are there more shipments to move; the unit transport price will typically be higher for smaller shipments. At the same time, stocks will remain small and inventory costs can be kept low. Larger shipments will have the reverse effect. Firms will then optimize their order size, trading off these two cost types. In the case of increased transport prices, the preference will be to order larger shipments, thereby reducing the effect of the cost increase on total costs. Figure 2.2 illustrates this in a stylized case. With transport costs increasing (right figure), the optimal shipment size increases as well.

The total logistics cost increase will be mitigated by this change in shipment size, leading to a lower total logistics cost increase than the original transport price increase (in the above example, a doubling of transport costs leads only to an increase of 50% of total costs). The repercussions of the tax will be that the amount of transport will be reduced (which is a benefit), but the total cost change that propagates into the system will be less than the price change. Similar trade-offs between the transport and inventory cost components can be found in the choice of mode of transport and vehicle type (Combes 2012) or distribution structures (Onstein et al. 2019). The real-world impact of such mechanism on transport flows is significant, as quantified by means of transport price elasticities in Davydenko (2015).

What are the relevant logistics decisions to consider? Riopel et al. (2005) list altogether 48 decisions in logistics management. Based on a condensed list (see Tavasszy et al. 2019), the below table explains their importance for the strategic routes for decarbonization, as mentioned in the introduction. The influence span of each decision area is limited. Interestingly, the choice of mode and transport efficiency are influenced by several decision areas, upstream of the transportation organization decisions. For example, during decisions on production and sourcing, geographical locations are determined which create transport demand – some of these will not allow access to rail and waterways networks. Also, sourcing decisions determine shipment sizes and frequencies, which will constrain the freedom to choose transport modes or to consolidate loads for higher efficiency (Table 2.2).

To conclude, logistics reorganization responses may involve many different decisions in several, vastly different areas of management. Ideally, all these areas

Table 2.2 Relevance of logistics decision areas for carbon pricing

Functional area	Carbon performance dimension and measure				
	Demand for transport (tonne-km)	Choice of mode (tonne-km per mode)	Transportation efficiency (vehicle/tonne)	Energy efficiency (J/vehicle km)	Carbon efficiency of energy (CO_2/J)
Sales and marketing	+	–	–	–	–
Production planning	+	+	–	–	–
Sourcing and material handling	+	+	+	–	–
Distribution channels	–	+	+	–	–
Inventory and packaging	+	+	+	–	–
Transportation organization	–	+	+	+	+

Legend: + influence, – little or no influence

would need to be addressed in order to cover the whole range of decarbonization routes.

In the next section, we introduce the six recent empirical cases in which logistics reorganization responses have been studied.

3 Overview of Case Studies, Foci and Methods

There is a large body of literature on the evaluation of the impact of carbon taxes. Our focus here is on predictive studies, not the normative approaches studying optimal configurations of supply chains (see, e.g. Das and Jharkharia 2018, for a review of these studies). Predictive evaluation studies have been presented at various levels of aggregation, including urban (Anand 2015; Waisman et al. 2013; Marcucci et al. 2018; Cheng et al. 2015), corridor (Winebrake et al. 2008; Zhang 2013), (inter)national (Piattelli et al. 2002; Raha et al. 2003; De Bok et al. 2019) and global level (Lee et al. 2013; Tavasszy et al. 2016; Halim et al. 2019). These used different types of models, falling largely into two categories: transport network models (including mode choice) and spatial/sectoral-economic interaction models (a gravity or CGE type approach). We are not aware of any case that considers decisions on production planning, distribution channel, inventories and fleet technology. The focus, therefore, lies on transport demand and a relatively rich view on transport organization. We distinguish between the following logistics choices:

Table 2.3 Case characteristics

Level	Local	Corridor	State	Continent	Global
Location	City of Rotterdam	Rhine delta	Netherlands	European Union	World
Policy type	Cap-and-trade	Container surcharge	Truck tolls	Truck tolls	Full internalization
Logistics decisions[a]	S, V, R	M, T, R	V, R	M, V, R	S, T, R
Dynamics	Event based	None	None	None	None
Modelling approach	Agent based simulation	Supernetwork choice	Discrete choice model	Discrete choice model	Linked SCGE and supernetwork
Source	Anand (2015)	Zhang et al. (2013)	De Bok et al. (2020)	Raha et al. (2003)	Tavasszy et al. (2016)

[a]**S**ourcing and sales, **M**ode of transport, **T**ranshipment location, **V**ehicle type, **R**oute choice

- Sourcing and sales of goods, resulting in trade (S)
- Mode of transport for inland modes (M)
- Transhipment terminals inland and for maritime traffic (T)
- Choice of vehicle type (truck size and engine type) (V)
- Route choice over the transport network (R)

Table 2.3 shows the detailed characteristics of the cases.

The policies considered are in one sense quite similar, as the main intervention in all cases is a tax per emitted tonne CO_2 applied to units of transport performance (tonne-km, vehicle-km or TEU-km). However, there were subtle differences in the mode of implementation. The case of the city of Rotterdam involved a limitation in the total volume of emission rights, to be distributed among all companies. Trading of these rights was made possible. The other cases did not involve a cap, but prices were varied in different ways. Case 2 varied prices to identify the tipping points in the system. Case 3 and 4 based the prices on the current markets for external effects, using principles of social marginal costs (case 3) or average market prices, insofar available (case 4). Case 5 extended well beyond carbon prices and included an internalization of other external costs as well including various emissions and safety costs of transport.

The cases differ in terms of the models used and the choices represented in these models. Case one was built on an agent-based model, where shopkeepers were the main decision-makers, deciding about order size (weighing transport costs against inventory costs) and type of vehicle (in effect, the carrier being either the legacy carrier or a common carrier using electric vehicles). Carriers would decide about their routes, combining shops served by round trips in the most efficient way possible. This is the only case in which dynamics was included, where the factor time in the model was driven by events of decision-making (which were assumed to have a fixed frequency) and by transport operations (the execution of which costs time). The second case uses a multimodal network (supernetwork) formulation to identify routing of flows in the hinterland of the maritime port of Rotterdam, the Netherlands, based on customer preferences for different types of services, by different modes. Case 3 concerns two applications of a very similar nature, one at national and one at international level. Both involved a relatively light form of truck pricing and involved responses for changes in vehicle types and routes driven. Case 4 also involved network choices (related to routes of transport including ports), but was linked to a model of trade relations between regions and sectors over the world.

In the next sections, we introduce each case and discuss the results of the applications.

4 Carbon Credit Points for City Logistics

The decarbonization solution for urban areas developed by Anand (2015) involves the case of a city trying to influence the external effects of freight distribution within its borders. It sets out a policy around a cap-and-trade policy for carbon emissions. Carbon permits are issued to all carriers entering the city; each trip into the city requires the use of one carbon credit point. As the number of points is limited (capped by the maximum volume of carbon emissions that the city wants to allow), the city government provides an alternative mode of shipping. An Urban Consolidation Centre (UCC), at the border of the inner city area, can be used to deposit freight destined for the city. From here, electric vehicles of a carrier concessioned by the city will take the freight to its destination. No petrol or diesel vehicles are allowed to enter the city, without accompanying carbon credits. As soon as all credits have been spent, remaining trips must be made via the UCC. Companies are allowed to trade carbon credit points. The price for trading in this scheme is set by the local government; the revenues are collected by government and recycled as a subsidy for the UCC. Credit points are perishable so that carriers cannot accumulate them (Fig. 2.3).

The situation was modelled using an agent-based formulation, where agents in a city (government, shopkeepers, carriers, UCC) made decisions. The decisions were aimed at achieving agent-specific objectives, were cyclical (i.e. reviewed with a fixed frequency) and, eventually, interdependent. The objective of government was to reduce the emission levels to a certain target level. Shopkeepers would aim to minimize their costs (demand being assumed fixed), just like carriers, while the UCC aimed to achieve a net financial result.

Decisions considered included the following (Fig. 2.4):

- Government: number and price of carbon credits, subsidy to UCC; monthly decision
- Shopkeepers: order size, carrier to use – daily decision
- Carriers: routing; daily
- UCC: price of transport with EV; monthly

Fig. 2.3 Lay-out of city logistics case

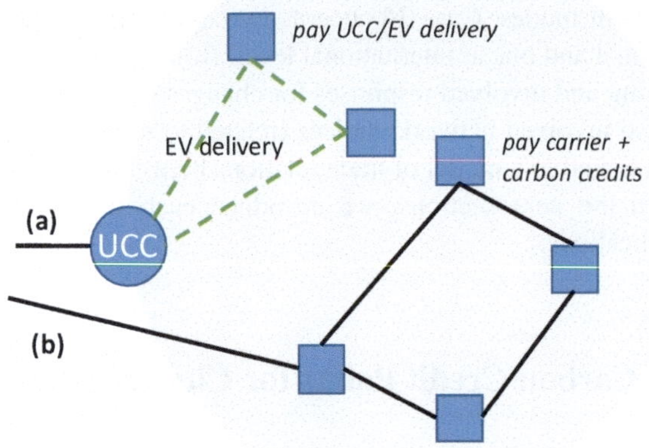

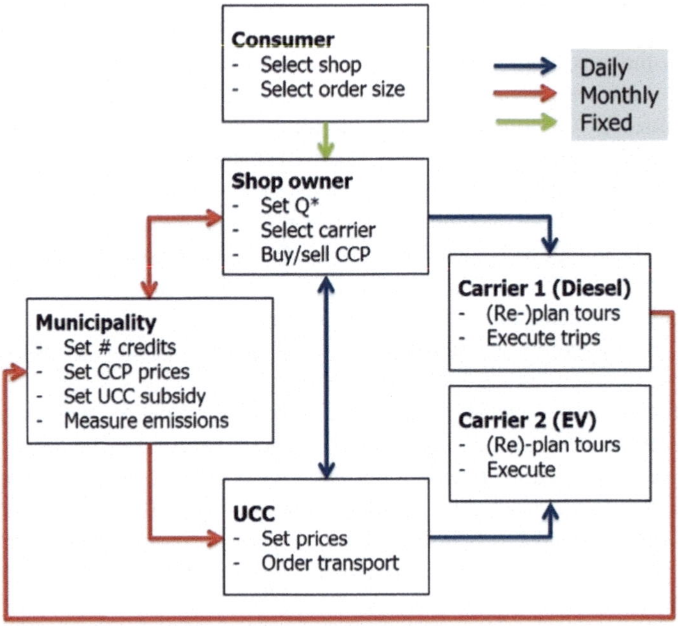

Fig. 2.4 Interrelated decisions of actors. Legend: review frequencies

One emergent property of the model that was of interest concerned the time that it would take for the target (a reduction in kilometres driven by conventional vehicles) to be reached. This required all the actors to align their decisions. The results were

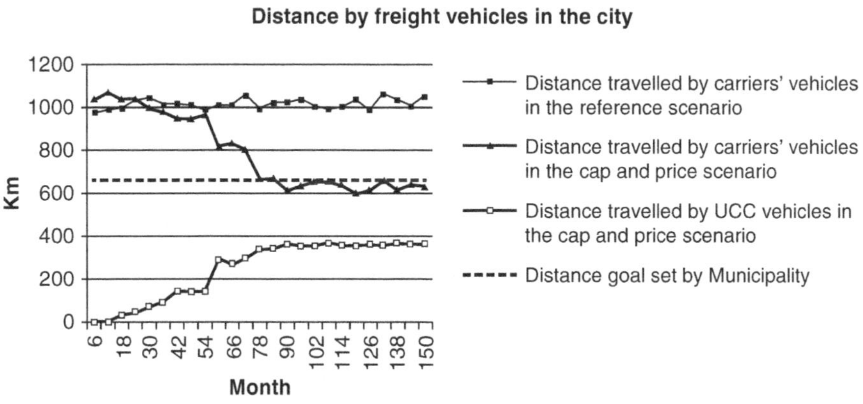

Fig. 2.5 Example of result of convergence to emission target. (Source: Anand 2015)

surprising in the sense that it would take years to reach the new equilibrium. Obviously, simple solutions could be found by increasing the frequency of decision-making of the government about CCP and UCC prices or even installing the equilibrium price at the outset. This would however force agents to have such a strong increase in costs at the short term that business continuity could be threatened. The main point, however, was that this model proved instrumental to predict the response time of the system (Fig. 2.5).

5 Hinterland Container Transport Corridor

The case of pricing in hinterland corridors was studied by Zhang (2013). She considered mode choice for the transport of containers between maritime ports and hinterland destinations. Depending on the destination, up to three alternative modes of transport are available: road, rail and waterways. Pricing policy was implemented as an additional variable transport charge for road transport, calculated via a carbon price. The main questions were (1) what price would be needed to create a significant shift of flows away from road transport to other modes and (2) which effect of CO_2 emission reduction could be achieved. Besides the decisions of shippers and forwarders about the mode of transport, also decisions on network design were considered. More specifically, terminal locations were regarded as flexible. As the policy would be implemented over the longer term, this strategic decision was included in the model.

The problem was modelled using a bi-level optimization approach, with service demand being determined by simulation of choices for the mode and route in a supernetwork (i.e. a fully connected multimodal network including terminals) and with network design being determined by an optimization model, solved using a

Fig. 2.6 Visualization of the supernetwork. (Zhang et al. 2013)

meta-heuristic based solution algorithm. The procedure is described in detail in Zhang (2013) (Fig. 2.6).

Subsequently, emission-based variable charging was introduced for all modes of transport. Users of the network would reconsider their choice of mode, based on the new price, and the network design would adapt to best serve demand. The emission charge was varied up to 1000 Euro/tCO_2. An optimum was found for a price of 150 Euro/tCO_2, where emissions had reduced by 20% compared to the base case. A higher price did not yield a lower emission level, implying that carriers had exhausted the opportunities to move to another mode of transport. Note that revenues were not recycled into the system.

The finding that emission-based pricing in hinterland freight systems can be reasonably effective up to a certain point has, to our knowledge, not been found elsewhere. Interestingly, the optimal price is not a very steep one. With an emission factor of 100 gCO_2eq/ton-km, a load of 10 tonne per container and a distance of 1000 km into the hinterland, the added price is 0,15 Euro per km., which is reasonably in line with the current policy proposals in the Netherlands – a mark-up of roughly 15% on the current trucking prices. Note that in this case, it seems safe to assume that the hinterland container transport market is decoupled from the logistics of the global supply chains using them, so that indirect efficiency impacts (changing shipment sizes, bundling of freight, adapted warehouse locations) can be neglected. As we will see in the case of national transport, however, this assumption is not a trivial one.

6 National and EU Road Networks

In this section, we report two transport model applications that were similar in nature but applied with slightly different models and at different spatial levels. Their comparison provides interesting insights for both cases. The first involves the large-scale network of the European Union. Although somewhat dated, the exercise is still unique in terms of the rationale of the policy and the scope of responses considered. The policy aimed to internalize external costs by means of a marginal external cost-based truck charging scheme. Marginal social costs were made dependent of the type of road, synthesizing external costs from various cost sources. Figure 2.7 shows the levels of external costs that coincided almost in a linear fashion with population density.

The model concerned the SCENES model of European passenger and freight transport (Raha et al. 2003). Besides the usual response of mode shift, also changes in vehicle type choices and trade relations were allowed. The effects of the truck charge differed by commodity type. All effects combined (mode, vehicle type, route) resulted in reductions in transport performance of up to 20% (Fig. 2.8). It is worthwhile to note that basic bulk products are relatively sensitive to transport price changes. Here, transport price has a relatively high share of the product price, and transport time is less important. Hence, shippers and carriers will be more inclined to respond and sacrifice service quality.

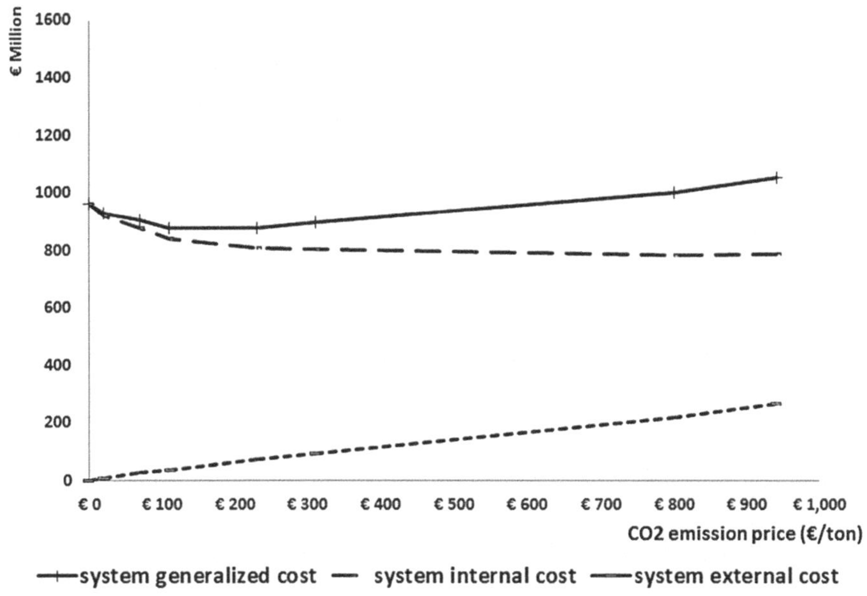

Fig. 2.7 Cost functions for internal, external and total costs. (Zhang et al. 2013)

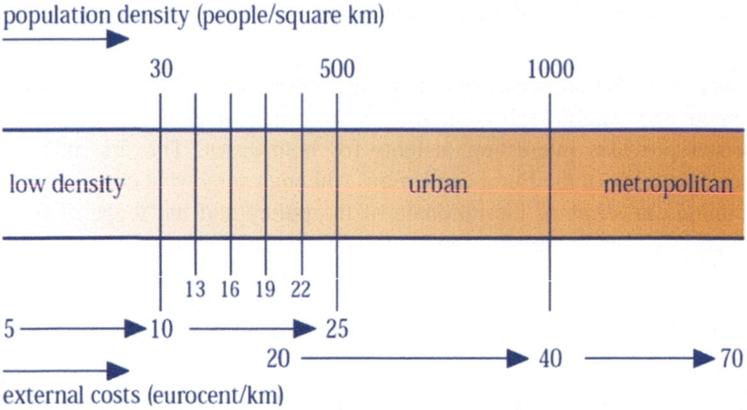

Fig. 2.8 External costs calculation related to urban density. (Raha et al. 2003)

A second case involves the most recent analyses of a Dutch variable road pricing scheme (MuConsult 2018). This study involved the use of expert judgement and the Dutch national freight transport model, BASGOED, for the assessment of impacts on transport flows. It was assumed that part of the charges would be absorbed by efficiency improvements in the logistics process, without detailed consideration of the background of these changes. The remaining part of the charges would propagate to the shippers and lead to changes in mode choice and trade patterns, similar as in the EU case. The idea that carriers would work more efficiently in response to the charges was based on the assumption that they would use larger trucks and transport bigger shipments. This conforms to the EOQ principle, which dictates that firms will order bigger shipments to reduce the costs of transport. The choice of shipment size, however, was not part of the model. In an additional exercise (De Bok et al. 2019), this effect was studied simultaneously with the choice of vehicle type. The finding was that the efficiency effect indeed existed, but was less prominent than assumed before. Increasing shipments create larger inventory costs for firms and will therefore restrain the increase of shipment size.

7 Global Supply Chains

At the global level, decarbonization through pricing is a major challenge, due to international competition. Choices made in response to pricing policy by consumers, producers and logistics service providers may lead to shifts in trade or transport activities from one country to another. A policy where countries price differently, and some not at all, may induce shifts of activities to so-called pollution havens. There have been few evaluations of a worldwide pricing policy on global trade and transport (Lee et al. 2013; Zeshan and Ko 2016); these have not considered impacts of a full internalization on both trade, economy and network flows. The policy

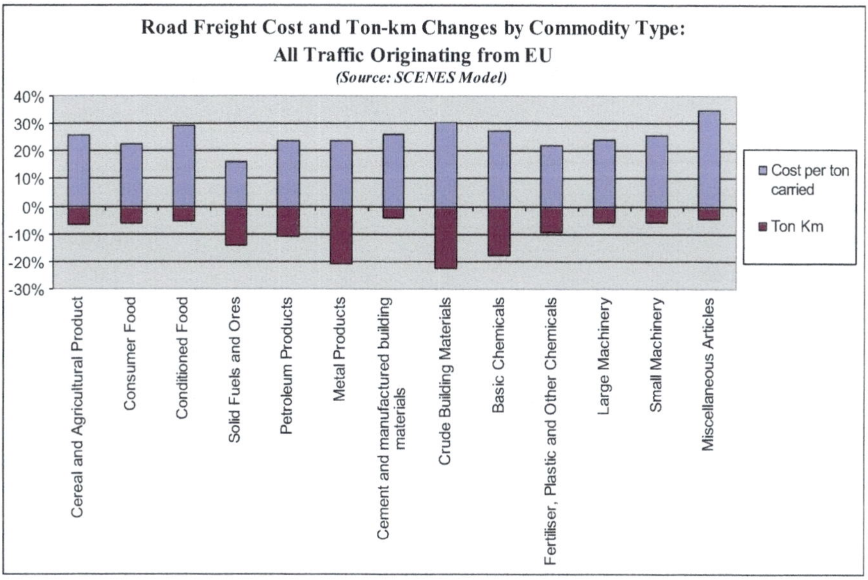

Fig. 2.9 Impacts of external costs based road user charges in the EU. (Raha et al. 2003)

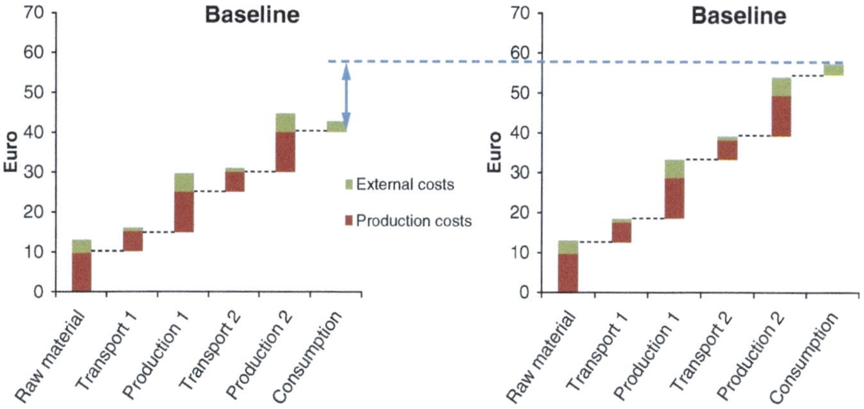

Fig. 2.10 Illustration of supply chain based internalization of external costs

evaluated in this case involved an internalization of all external costs in supply chains and is reported in more detail in Tavasszy et al. (2016). External costs associated with various types of health effects were included, going far beyond carbon emissions (environmental externalities and social externalities including noise, congestion and traffic safety). Figure 2.9 shows the principle of adding external costs to product prices in the entire supply chain. Eventually, consumers will be charged the full external costs (Fig. 2.10).

Table 2.4 Impact of world trade and transport of internalization of external costs

Sectoral added value (USD)	Impact (2040)
Agriculture	−4.2%
Food and feed	−0.2%
Solid fuels	−1,1%
Oil	−0.9%
Ores and scrap	−0.6%
Metals	−0.3%
Building materials	−1.0%
Fertilizers	−0.9%
Chemicals	−0.7%
Manufacturing	−0.3%
Overall impact on global trade (USD)	**−0.9%**
Impact on Rotterdam's throughput (TEU/y)	**−0.5%**
Impact on CO_2 emission (ton)	**−27%**

To evaluate changes in world trade, a macro level model of global supply chains was used, in the form of a Spatial Computable General Equilibrium (SCGE) model of the global economy. The SCGE model EXIOMOD (Ivanova 2014) was linked to a network model (Tavasszy et al. 2011) to understand changes in the global routing of transport flows. The internalization scenario produced a new global economic equilibrium for production, consumption and trade, and the network assignment showed how these new trade flows would affect main maritime ports in the network, such as the port of Rotterdam. Table 2.2 summarizes the results of the calculations for the Netherlands.

The trade of traditionally environment unfriendly sectors such as agriculture (particularly meat production), non-renewable energy and building materials would suffer most, but only at a relatively modest rate of up to 4.2% aggregate growth of value added over a period of more than 25 years. The overall impact on global trade would amount to less than 1% over this period, and port throughput would be reduced only by 0.5%. Emissions would be 27% lower, however, than the baseline volumes in 2040 (Table 2.4).

Although this may not seem a big change compared to the targets set for global emission reduction, one should bear in mind that the reduction was achieved with current market prices for all external costs. An interesting finding, therefore, is that the long supply chains, in effect part of an extensive network of sectors within the economy, are able to adjust fairly effectively. While the many small adjustments all absorb part of the price increase, the net increase at the end of the chain is small. The additive effect of emission reductions, in line with the internalization concept, is high, however.

8 Synthesis and Lessons from Cases

The main lessons across the cases concern a number of topics; we discuss these below:

- Importance of understanding logistics decisions
- Results suggest trade-off between acceptance and impact
- Identified gaps in research

8.1 *Importance of Understanding Logistics Decision-Making*

The cases presented vary widely in terms of the logistics decisions considered. The exact influence of every decision on the entire impact chain – between truck costs and truck volumes – is still largely uncharted territory. Systematic reviews of price elasticities of road transport flows do exist (de Jong et al. 2010; Beuthe et al. 2014), but only give a partial account of decisions, as focusing on mode choice-based price sensitivity. As we see from the national cases, however, it is important to include aspects such as inventories and logistics efficiency. Also, the choice of vehicle type is relevant. This series of logistics decisions can act as a buffer in the impact chain, dampening the effect of price increases. Only recently, new elasticities on distribution channel-redesign have been investigated (Davydenko 2015). For many other logistics reorganization responses, the impact at a macro level is unclear. A separate issue concerns the degree to which carriers van transfer the price increase to their clients (Holguín-Veras et al. 2015). More work is needed that describes the behaviour of a population of consecutive logistics decision-makers.

One consequence of the long chains of decision throughout the supply chain is that price increases will be absorbed to a large extent by the time they reach the consumer. The global case shows clearly that the intersectoral networks allow many options to replace polluting products and services by those that pollute less – to an extent that the final effects will hardly be felt. This effect also appeared in another comparable study (Lee et al. 2013).

8.2 *Trade-Off Between Acceptance and Impact*

All studies indicate that current market prices for external costs are too low for internalization to work through in significant reduced carbon emissions. The corridor case study shows that carbon prices need to be an order of magnitude above the prices of this decade (at current, European Emission Allowance prices are varying roughly between 5 and 30 Euro per tonne). Above levels of 150 Euro/tCO_2, transport activity proved relatively unresponsive again, suggesting that the 20% reduction is the feasible potential for a hinterland network with sufficient options available over water and by rail. The current market prices, roughly translated into the charge calculations in the Dutch and the global case, had a minor impact. The EU variable charge was based on a systematic account of total external cost and had a significantly higher impact. Clearly, if one wants to arrive anywhere near a contribution to global emission targets, higher charges have to be considered than what is now considered politically feasible.

8.3 Gaps in Studies

The studies described here are representative of the broader extant literature, in terms of the scope of decisions considered, but relatively rich compared to mainstream research. We have shown that it matters for the result of impact assessments, which reorganization responses are included in studies. A number of important gaps in modelling would need to be addressed. Firstly, many decisions that could have a mitigating effect on industry responsiveness are still missing in most studies, like distribution chain design, outsourcing of logistics services and shipment sizes. Secondly, more insight is needed in the distributional effects of policies towards specific sectors and regions of the world. These may be large and may require additional policies to promote acceptance of decarbonization. Thirdly, the calculations seem to indicate that the level of decarbonization needed in the sector of around 80–90% will not be achieved by taxing current technologies. None of the studies presented here and, to our knowledge, no other work in the extant literature arrives at decarbonization results of this scale. The question, therefore, how a drastic decarbonization can be achieved, is not answered. Finally, the issue of dynamics seems to be a weak spot in the literature – while assumptions about dynamics are essential to predict times to impact of policy measures, we are essentially still in the dark. We note, finally, that a comprehensive literature review would be needed to confirm the above issues that we have only been able to touch upon briefly with these cases.

9 Concluding Remarks

The purpose of this chapter is to explore ways in which logistics decisions can be taken into account in evaluations of decarbonization policies. To this end, we reviewed and compared several cases of quantitative policy evaluation, all implementing a form of taxation on transport, with the aim to reduce external effects. The cases underline a number of general lessons. Firstly, logistics decision-making is important to be understood as they determine the reorganization response of companies. The current literature on logistics decision-making is too narrow in scope to cover the necessary range of relevant choices and their properties. Additional research is needed to be able to predict impacts of carbon taxes. Typical balancing mechanisms in logistics such as the trade-off between transport and inventory costs, and empirically difficult questions such as decision dynamics, should be prioritized for research. Secondly, from a policy perspective, the level of carbon prices necessary to push emissions in the system downward are an order of magnitude higher than the current market prices. For policy makers, this suggests that attention is needed for the public acceptability of effective carbon taxes, or other non-fiscal measures aimed at an equivalent impact. Thirdly, as logistics decisions are multi-faceted and supply chains include many companies, the impact chain between tax measures and emission volumes is long. This implies that the ultimate impact on consumers of carbon taxes can be small, while the added effect of emission savings is expected to be relatively large.

References

Anand, N. (2015). *An agent based modelling approach for multi-stakeholder analysis of city logistics solutions* (PhD thesis). Delft: TU Delft.

Beuthe, M., Jourquin, B., & Urbain, N. (2014). Estimating freight transport price elasticity in multi-mode studies: A review and additional results from a multimodal network model. *Transport Reviews, 34*(5), 626–644.

Cheng, Y. H., Chang, Y. H., & Lu, I. J. (2015). Urban transportation energy and carbon dioxide emission reduction strategies. *Applied Energy, 157*, 953–973.

Combes, F. (2012). Empirical evaluation of economic order quantity model for choice of shipment size in freight transport. *Transportation Research Record, 2269*(1), 92–98.

Das, C., & Jharkharia, S. (2018). Low carbon supply chain: A state-of-the-art literature review. *Journal of Manufacturing Technology Management, 29*(2), 398–428.

Davydenko, I. Y. (2015). *Logistics chains in freight transport modelling* (PhD dissertation). Delft: TU Delft.

De Bok, M., Bal, I., Tavasszy, L., Tillema, T., & Francke, J. (2019). *Exploring the impacts of an emission based truck charge in The Netherlands* (Working Paper TU Delft, mimeo).

De Bok, M., Bal, I., Tavasszy, L., & Tillema, T. (2020). Exploring the impacts of an emission based truck charge in the Netherlands. *Case Studies on Transport Policy, 8*(3), 887–894.

de Jong, G. C., Schroten, A., van Essen, H., Otten, M., & Bucci, P. (2010). The price sensitivity of road freight transport: A review of elasticities. In *Applied transport economics, a management and policy perspective*. Antwerp: De Boeck.

Halim, R. A., Smith, T., & Englert, D. P. (2019). *Understanding the economic impacts of greenhouse gas mitigation policies on shipping: What is the state of the art of current modeling approaches?* The World Bank.

Holguín-Veras, J., Aros-Vera, F., & Browne, M. (2015). Agent interactions and the response of supply chains to pricing and incentives. *Economics of Transportation, 4*(3), 147–155.

ICF & HLB. (2002). Economic effects of transportation: The freight story, final report. In *Fairfax*. ICF Consulting: VA.

Ivanova, O. (2014). Modelling inter-regional freight demand with input–output, gravity and SCGE methodologies. In L. A. Tavasszy & G. de Jong (Eds.), *Modelling freight transport* (pp. 13–42). Elsevier.

Kaya, Y., & Yokoburi, K. (1997). *Environment, energy, and economy: strategies for sustainability*. Tokyo: United Nations University Press. ISBN 9280809113.

Lee, T. C., Chang, Y. T., & Lee, P. T. (2013). Economy-wide impact analysis of a carbon tax on international container shipping. *Transportation Research Part A: Policy and Practice, 58*, 87–102.

Marcucci, E., Gatta, V., & Le Pira, M. (2018). Gamification design to foster stakeholder engagement and behavior change: An application to urban freight transport. *Transportation Research Part A: Policy and Practice, 118*, 119–132.

McKinnon, A. (2018). *Decarbonizing logistics: Distributing goods in a low carbon world*. Kogan Page Publishers.

MuConsult. (2018). *Effectstudies vrachtwagenheffing (Impact study of truck charges), Study for the Netherlands Ministry of Transport and Public Works*. Amersfoort: MuConsult.

Onstein, A. T., Tavasszy, L. A., & van Damme, D. A. (2019). Factors determining distribution structure decisions in logistics: A literature review and research agenda. *Transport Reviews, 39* (2), 243–260.

Piattelli, M. L., Cuneo, M. A., Bianchi, N. P., & Soncin, G. (2002). The control of goods transportation growth by modal share re-planning: the role of a carbon tax. *System Dynamics Review: The Journal of the System Dynamics Society, 18*(1), 47–69.

Raha, N., Jin, Y., Rustenburg, M., Tavasszy, L. A. (2003). *The impacts of pricing of truck transport in the EU*. Proceedings European Transport Conference, available at https://aetransport.org/past-etc-papers

Riopel, D., Langevin, A., & Campbell, J. F. (2005). The network of logistics decisions. In *Logistics systems: Design and optimization* (pp. 1–38). Boston, MA: Springer.

Tavasszy, L., Minderhoud, M., Perrin, J. F., & Notteboom, T. (2011). A strategic network choice model for global container flows: specification, estimation and application. *Journal of Transport Geography, 19*(6), 1163–1172.

Tavasszy, L., Harmsen, J., Ivanova, O., & Bulavskaya, T. (2016). Effect of a Full Internalization of External Costs of Global Supply Chains on Production, Trade and Transport. *Towards Innovative Freight and Logistics, 2*, 337–351.

Tavasszy, L., de Bok, M., Alimoradi, Z., & Rezaei, J. (2019). Logistics decisions in descriptive freight transportation models: A review. Journal of Supply Chain Management Science.

UNEP. (2016). *The Emissions Gap Report 2016*. Nairobi: United Nations Environment Programme (UNEP).

Waisman, H. D., Guivarch, C., & Lecocq, F. (2013). The transportation sector and low-carbon growth pathways: modelling urban, infrastructure, and spatial determinants of mobility. *Climate Policy, 13*(sup01), 106–129.

Winebrake, J. J., Corbett, J. J., Falzarano, A., Hawker, J. S., Korfmacher, K., Ketha, S., & Zilora, S. (2008). Assessing energy, environmental, and economic tradeoffs in intermodal freight transportation. *Journal of the Air & Waste Management Association, 58*(8), 1004–1013.

Zeshan, M., & Ko, J. H. (2016). An Analysis of the Economic Impact of Implementing the INDCs of the 2015 Paris climate conference: A CGE approach. *International Trade Research, 2*(4), 81–110.

Zhang, M. (2013). *A freight transport model for integrated network, service, and policy design* (PhD thesis). Delft: TU Delft.

Zhang, M., Wiegmans, B., & Tavasszy, L. (2013). Optimization of multimodal networks including environmental costs: a model and findings for transport policy. *Computers in industry, 64*(2), 136–145.

Chapter 3
The Impact of Collaborative Scheduling and Routing for Interconnected Logistics: A European Case Study

Sh. Sharif Azadeh, Y. Maknoon, J. H. Chen, and M. Bierlaire

Abstract Interconnected logistics system can play an important role towards having a more sustainable green freight transport. Recently, after introducing the concept of Physical Internet (PI), researchers have started to explore the opportunities and challenges that a collaborative and interconnected network could create in different aspects of the supply chain. In this research, we study the last mile delivery as well as vehicle dispatching problems under the assumptions of collaborative supply chain networks while assuming that modularized boxes are applied inside the network from the provider to the final customer. Our research aims at proposing a more efficient resource planning with the minimal number of empty vehicle movements running on roads that ultimately leads to decrease carbon dioxide emission. The assumptions have been tested and verified using real data coming from a major retail company in Europe.

Keywords Collaborative scheduling · Routing · Interconnected logistics · Physical Internet · Last mile delivery · Vehicle dispatching

Sh. Sharif Azadeh (✉)
Earsmus University Rotterdam, Rotterdam, The Netherlands
e-mail: sharifazadeh@ese.eur.nl

Y. Maknoon
Delft University of Technology, Delft, The Netherlands
e-mail: m.y.maknoon@tudelft.nl

J. H. Chen
Jiao Tong University, Shanghai, China
e-mail: jianghang.chen@sjtu.edu.cn

M. Bierlaire
École Polytechnique Fédérale de Lausanne, Lausanne, Switzerland
e-mail: michel.bierlaire@epfl.ch

J. Rezaei (ed.), *Strategic Decision Making for Sustainable Management of Industrial Networks*, Greening of Industry Networks Studies 8,
https://doi.org/10.1007/978-3-030-55385-2_3

1 Introduction

In 2011, the European Commission published a white paper in which it formulated the long-term ambition to reduce greenhouse gas emissions from transport by at least 60% by 2050 compared to 1990. The emissions increased by 26% compared with 1990 levels. This increase comes despite past improvements in the efficiency of transport and is broadly in line with increases in the level of economic activity as measured by gross domestic product (GDP) as well as increases in demand for transport. Road transport accounts for 72% of total greenhouse gas emissions of the sector. Further increasing the efficiency of the logistic system in road transport will play a key role in limiting the increase of road transport emissions.

Nevertheless, total transport demand is predicted to continue growing during the 2020–2030 period in line with 2010–2020 patterns (1.5% for freight transport (tonne km)) and at lower rates between 2030 and 2050 (0.8% for freight transport).

Integrated measures addressing both production and consumption would therefore be needed in the long run in order to reduce the greenhouse gas emissions from transport by 60% by 2050 (European Environmental Agency 2018).

In order to make a better use out of logistics resources and to exploit synergies between different distribution service providers, the concept of Physical Internet (PI) and interconnected networks were introduced (Montreuil 2010). PI proposes to use a new framework of interconnected logistics especially designed for resource sharing, real-time identification, and routing through open facilities to use transport infrastructure more efficiently and reduce environmental impact.

Within this framework, all products are encapsulated in smart, modularized, ecofriendly and standard boxes loaded and then handled, stored, and transported through shared facilities and across open networks.

There are two significant characteristics of the Physical Internet: encapsulation and collaboration.

Encapsulation: The Physical Internet does not manipulate physical goods directly. Instead, it manipulates exclusively containers that are explicitly designed for the Physical Internet and that encapsulate physical goods within them (Montreuil 2011). These dedicated containers for the Physical Internet have modular dimensions and standardized interfaces for handling and communication.

Collaboration: The Physical Internet provides universal and standardized interfaces and protocols to reduce the frictions in supply chain horizontal collaboration. For any logistics services providers, as long as they accept the operational protocols to handle, move, store, transport, and use the Physical Internet containers, they become the members, beneficiaries, and collaborators in the Physical Internet despite their potential competitive relationships in businesses.

The main contributions of our research are to address two major problems in supply chain management under PI assumption when modularized boxes are used as containers for different products: (1) last mile delivery integrated with bin packing problem and (2) vehicle dispatching problem.

1.1 Sustainable Last Mile Delivery

Even though passenger mobility has received considerable attention in the literature and in practice in the recent years (Gentile and Noekel 2016; Alonso-Mora et al. 2017), other contributions to new research and technology are found for the modeling of last mile delivery within urban areas. Applications can be found in different industries to tackle the issues around last mile delivery in urban areas. To name a few, we can mention DPD (https://www.dpd.com/), Green Link (http://green-link.co.uk) or self-service parcel stations from DHL Packstation, LaPoste Pickup Station, etc.

The most commonly used vehicles for deliveries in the last mile delivery (including the request made via online shopping) are vans or trucks. The increase in e-commerce and related deliveries in cities is contributing to the increase in van traffic resulting in more pollution. For example, in the UK, these vehicles are responsible for 15% of total kilometers traveled on roads in 2015 compared to 10% in 1993 (Bates et al. 2018). In addition, these vehicles have contributed in 13.3 million tonnes of CO_2 equivalent to emissions in 2014 (Zanni and Bristow 2010). In this research, we focus on two aspects of the urban logistics systems in order to reduce the number of necessary vehicles and kilometers traveled by them in the network. In addition, we aim at shed light on how the available space inside the vehicles can be used more efficiently to avoid circulating empty vehicles on roads. Both topics are defined within the framework of Physical Internet.

1.2 Last Mile Delivery and Bin Packing Problem

The last mile delivery problem has been recognized as one of the most expensive, least efficient and one of the main responsible to polluting inside the supply chain networks. In urban areas, traffic infrastructure is used for the purpose of delivering goods that results in traffic jams (Ehmke 2012). Not having a good planning system for the last mile delivery causes heavier traffic that affects service quality and the final cost (Eglese 2006). The body of literature is quite rich when it comes to the last mile delivery. Here, we briefly mention the most relevant papers to our work.

In Gendreau et al. (2006), the authors propose a Tabu search in order to solve the vehicle routing problem with capacity and route length restrictions. The Tabu search consists of examining successive neighbors of a solution and selects the best. The authors use a generalized insertion procedure that repeatedly removes a vertex (which represents a customer) from its current route and reinsert it into another route. This is the neighborhood of a solution. In order to avoid cycling, solutions that were recently examined are forbidden and inserted in a constantly updated Tabu list.

In Bortfeldt (2012), the author presents a hybrid algorithm for the three-dimensional loading capacitated vehicle routing problem. It includes a Tabu search algorithm for the routing part and a tree search algorithm for packing boxes into vehicles. The Tabu search starts with a randomly generated solution. Then, for each route found by the Tabu search, the tree search algorithm tries to generate a packing plan where all boxes are placed correctly. Each node of the tree has three elements: a partial solution of placements, a set of free boxes that must be placed, and a list of potential placements. The algorithm tries to add placements for each free box until all are placed, or a time limit has been reached, or one box has no possible placement.

In Massen et al. (2012), for a similar problem, it uses an ant colony algorithm combined with a column generation algorithm which is used to solve large linear programming programs. Column generation generates only the variables which can potentially improve the objective function.

Only very small instances for the bin packing problem have been solved to optimality. Some exact methods were proposed by Martello et al. (2000). For bigger instances, only heuristic methods have been developed. In Hifi et al. (2010), the authors consider the assignment of items to identical bins. The packings have to be feasible, and their aim is to minimize the number of bins needed. n items characterized by a width w_i, a height h_i, and a depth d_i ($i = 1,2,\ldots\ldots,n$) are put in identical bins with width W, height H, and depth D. By using integer linear programming, they are able to find solutions for the bin packing. The constraints that must be satisfied are expressed as inequalities.

In Levine and Ducatelle (2004), an ant colony optimization is presented, in order to solve bin packing and cutting stocks problems. It is inspired by the capability of ants to find the shortest path between their nest and a food location by using pheromone trails.

The authors in Fanslau and Bortfeldt (2010) present a tree search algorithm to solve the 3D container loading problem for weakly or strongly heterogeneous items (i.e., same or different dimensions). They fill a container by adding blocks which are arrangements of one or more oriented boxes (items). The blocks are placed in residual spaces. In order to find the best block for a residual space, a tree search is used.

Our research is closest to the works of Gendreau et al. (2006) and Massen et al. (2012); however, our research in this chapter focuses on the impact of horizontal collaboration for the last mile delivery in the context of Physical Internet. This leads to better usage of capacities and reducing the operational cost as well as the number of vehicles required to deliver products.

1.3 The Vehicle Dispatching Problem

Physical Internet hubs are the places where modularized containers are sorted, assembled, and packed into vehicles. According to the concept of the Physical Internet, the short-range transportation is encouraged, which means that if possible, all the transportation activities are ideal to be limited between a Physical Internet hub and one of its neighboring hubs. The rational of such a recommendation is to

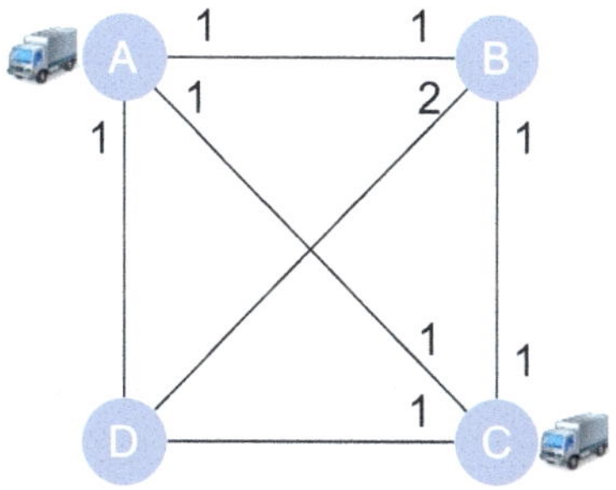

Fig. 3.1 An example to illustrate the concept of the vehicle dispatching problem

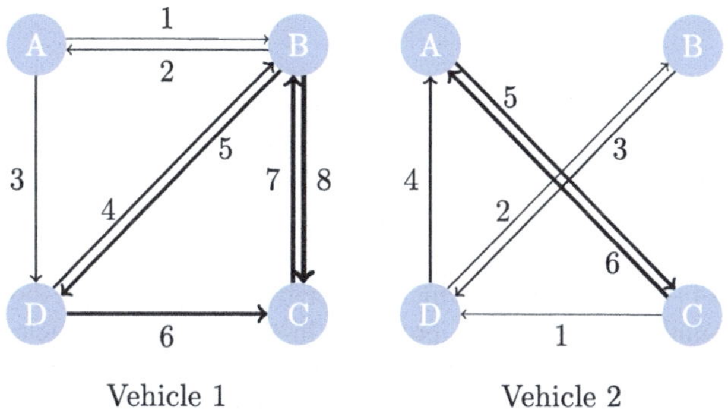

Fig. 3.2 A feasible solution to the example in Fig. 3.1

maximize the overall social benefit of truck drivers so that they are able to come back home after their daily jobs. In this section, we will introduce an optimization problem called the vehicle dispatching problem inside the PI context. Here, in transportation demands between any two linked hubs (in terms of how many trailers of modularized boxes to ship) and the current locations of all the available vehicles designated for the network (assuming all vehicles are homogeneous with the transportation capacity equal to 1 trailer), the decision-makers need to design a vehicle dispatching plan for each vehicle such that all the transportation demands are satisfied and the entire vehicle traveling cost is minimized. For example, in Fig. 3.1, there are four PI hubs, i.e., A, B, C, and D. The traveling distance between any two connected hubs is 1 and the transportation demands are listed in the figure. For instance, there is one trailer to be transported from hubs A to B and two trailers to be transported from hubs B to D. Initially, there are two vehicles positioned at hubs A and C, respectively.

Figures 3.2 and 3.3 depict two different dispatching plans for the vehicle dispatching problem shown in Fig. 3.1. In the solution shown in Fig. 3.2, vehicle 1 initially resided in Physical Internet hub A takes the path $A \Rightarrow B \Rightarrow A \Rightarrow D \rightarrow B \Rightarrow D \rightarrow C \Rightarrow B \Rightarrow C$ with traveling cost 7 while the other vehicle initially positioned at hub C takes the path $C \Rightarrow D \rightarrow B \Rightarrow D \rightarrow A \Rightarrow C \Rightarrow A$ with traveling cost equal to 6. Note that the arcs in the walks symbolized as $\Rightarrow$ indicate that a vehicle is fully

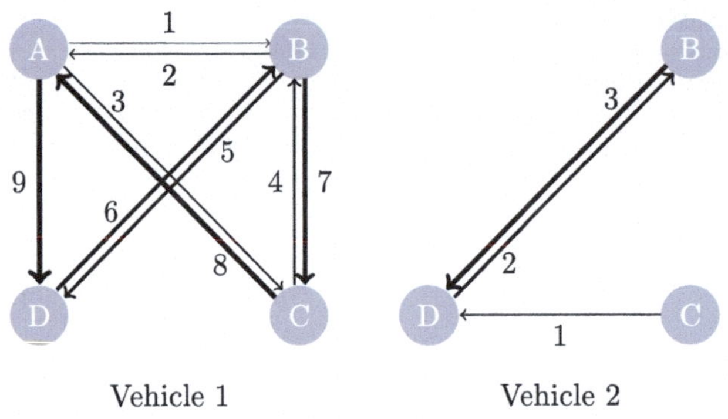

Fig. 3.3 An optimal solution to the example in Fig. 3.1

loaded. Oppositely, the arcs with symbol $\rightarrow$ represent the empty moves of a vehicle. Therefore, it can be observed that in the solution shown in Fig. 3.2, all the transportation demands can be fulfilled and the total traveling cost for such a dispatching plan is equal to 14 with 4 empty vehicle movements. Figure 3.3 shows a better dispatching solution for the same problem. In this solution, vehicle 1 takes the path $A \Rightarrow B \Rightarrow A \Rightarrow C \Rightarrow B \Rightarrow D \rightarrow B \Rightarrow C \Rightarrow A \Rightarrow D$, and vehicle 2 takes the path $C \Rightarrow D \rightarrow B \Rightarrow D$. Apparently, the total traveling cost for this solution is 12, and the number of empty vehicle movements has been reduced to 2 instead of 4 compared to the solution shown in Fig. 3.2. Actually, the solution shown in Fig. 3.3 is an optimal one for the vehicle dispatching problem in Fig. 3.1. The purpose of the proposed vehicle dispatching problem for the Physical Internet is very meaningful since it aims to seek the best resource dispatching plan with the minimal number of empty vehicle movements and thus ultimately the least carbon dioxide emission.

The remainder of this chapter is as follows: in Sect. 2, we introduce the integrated last mile delivery and 3D bin packing problem followed by the numerical results of the model presented in Sect. 3. In Sect. 4, the vehicle dispatching problem and its associated computational results are depicted. We conclude the chapter in Sect. 5 by also shortly discussing about potential future research avenues.

2 Last Mile and Bin Packing Problem

The last mile problem in the Physical Internet aims to deal with the final deliveries of the orders (encapsulated in modularized boxes) from hubs to its served customers. As illustrated in Fig. 3.4, there are five customers (A, B, C, D, and E) in the region that is served by a Physical Internet hub. Each customer has a list of modularized boxes (three types in this example, colored by red, green, and yellow, respectively) that should be delivered by a truck originally located at the hub. Taking customer E as example, 1 red box, 1 green box, and 2 yellow boxes are ordered, and these boxes

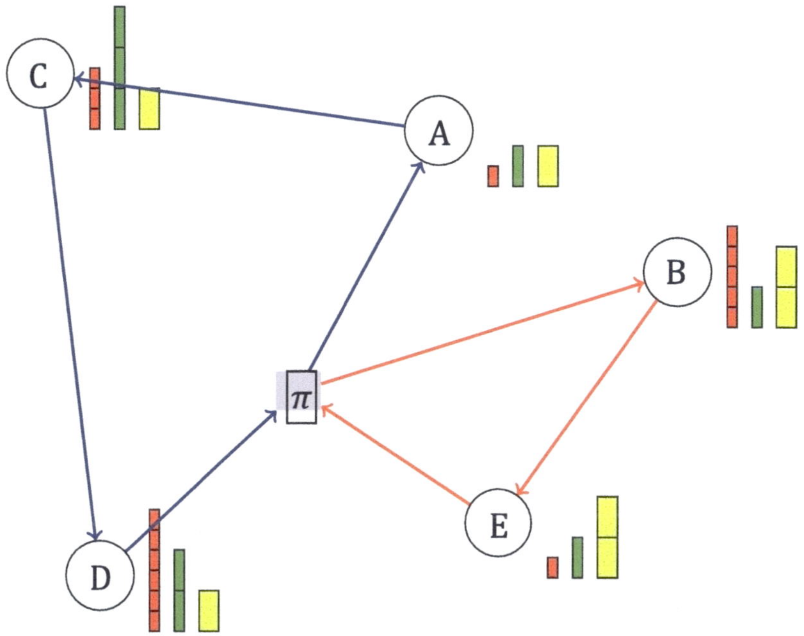

Fig. 3.4 The last mile delivery problem

should be delivered in batch. Thus, the major decisions for the operator of the PI hub are as follows:

1. How many vehicles should be used to fulfill the delivery task?
2. For each vehicle, which customers should be served?
3. For each vehicle, after assigning the served customers, which visiting sequence should be adopted by the driver in order to minimize the total traveling distance?

In this vehicle routing problem (VRP), each vehicle has a weight and volume capacity for its trailer. However, compared to the traditional capacitated VRP, here, the dimensions of the modularized boxes should also be taken into account when checking the feasibility of a routing plan for a vehicle. Figure 3.5 shows the necessity of the consideration in two-dimensional space. For example, let the internal size of a trailer be rectangle *ABCD*, and the hatched area is the occupied space by other modularized boxes. Box *FKGH* is the next container to be packed into the trailer which has the volume exactly equal to the current available space (i.e., the rectangle *FECD*). Although the overall sum of volume does not exceed the trailer's capacity, the box *FKGH* could not be packed into the trailer even when we allow the rotation of the box.

Secondly, the visiting sequence of clients for a vehicle poses the constraints on the packing sequence of the corresponding modularized boxes ordered by the customers. Such a consideration is referred to be as the rule of "Last-In, First-Out." For example, as shown in Fig. 3.4, there is one vehicle whose visiting

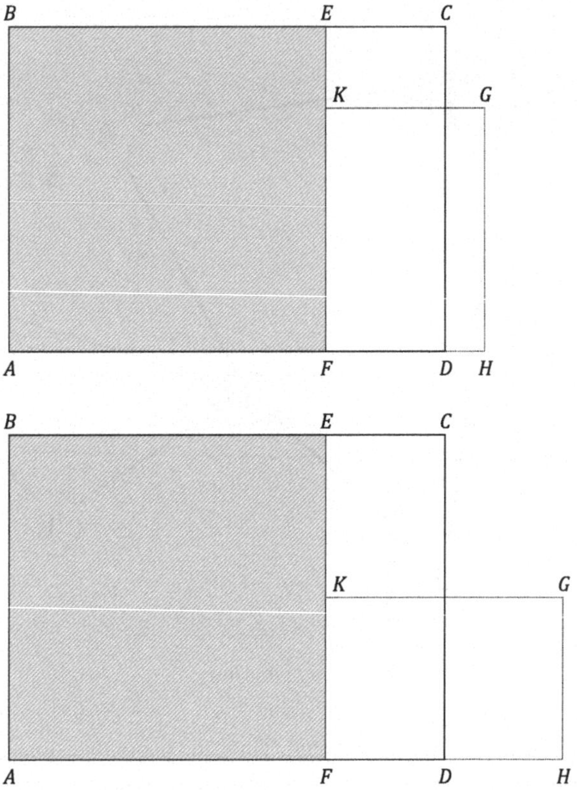

Fig. 3.5 The necessity to consider the dimensions of the modularized boxes in packing

sequence is $\pi \rightarrow B \rightarrow E \rightarrow \pi$. Therefore, the modularized boxes ordered by customer B should be packed in a way that when the vehicle arrives at B, all the boxes cannot be hindered by other boxes to the downstream customers when discharging. We refer the proposed problem for the Physical Internet as the vehicle routing problem coupling with 3D bin packing.

2.1 Problem Description

In this subsection, we provide the detailed problem description of the problem by introducing the input information as well as the general objective and constraints.

Input Parameters

- The geographical information of all the customers for the last mile delivery
- The modularized boxes ordered by all the customers (i.e., numbers, dimensions, and weights)

- The weight/volume capacities of the vehicles (assume that all the vehicles are homogeneous in this study)

3 Objective

The main objective of the proposed problem is to find out the best routing strategy such that the total traveling distance of all the vehicles are minimized (while reducing the total number of vehicles used for the delivery).

Constraints

1. The weight/volume capacity for all the vehicles cannot be exceeded.
2. All the modularized boxes assigned to one vehicle should be able to be completely packed into the assigned vehicle.
3. The rule of "Last-In, First-Out" should be respected.
4. All the customers can only be visited once. That is, all the ordered modularized boxes should be arrived at the location of the customers in a batch mode.

In the following section, we explain the algorithms used to solve the above problem.

3.1 Resolution Approach

Both the VRP and the 3D bin packing problems are NP-hard (Pinedo, 2012). In our case, their combination is also NP-hard since both of its subproblems are NP-hard. Consequently, heuristic methods are required to solve the problem. In Gendreau et al. (2006) and Massen et al. (2012), the authors used metaheuristic methods such as Tabu search and ant colony optimization to reach near optimal solutions. Such metaheuristics use randomness to explore better solutions. However, one of the accompanying counter effects of such approaches is that different runs of the same algorithm may result in different final solutions. Since, as mentioned before, the main purpose of this last mile problem is to analyze the potential of the horizontal collaboration, to prevent the generation of inconsistent results affected by randomness, the solving algorithms for the proposed last mile problem are deterministic rule-based heuristics.

The master problem of the proposed last mile problem is dedicated to vehicle routing, and three-dimensional bin packing problem plays the role of a side constraint. Our proposed algorithm introduces an insertion heuristic to solve the VRP as master problem. As commented by Campbell and Savelsbergh (2004), insertion heuristics have proven to be popular methods for solving a variety of vehicle routing problems due to their computational efficiency and the ability to be easily extended to handle complicated constraints (e.g., the three-dimensional bin packing in this study).

Algorithm 1 *

Algorithmic framework to solve the last mile problem

```
  N = set of unassigned customers;
  R = set of routes, always contains the empty route, initially contains only the empty
  route; while N , ∅ do c* = ∞; for j ∈ N do
          for r ∈ R do
              for (i − 1, i) ∈ r do
                          if BinPackingFeasible(r, i, j) and Cost(i, j) < c* then
                      r* = r; i* =
                      i;
                      j* = j;
                      c* = Cost(i, j);
                  end if
              end for
          end for
     end for
     Insert (i*,j*);
     N = N \ j* ;
     Update(r*);
 end while
```

In the framework above, *BinPackingFeasible(r,i, j)* is a function to check whether the insertion of customer j between $(i - 1)$ and i in the route r is feasible for three-dimensional bin packing such as the non-overlapping of modularized boxes and the Last-In, First-Out requirement. We proposed two methods to evaluate the value of the function *BinPackingFeasible(r,i, j)*. The first one is based on the technique of constraint programming. Compared to traditional mathematical optimization, usually in constraint programming, the optimal solutions are not very important since the major task is to seek feasible solutions which satisfy all the constraints. However, in the three-dimensional bin packing problem, since we must obey the laws of gravity and cannot allow "floating boxes," we try to minimize the sum of y_is instead.

Algorithm 2 *

Constraint programming to evaluate BinPackingFeasible(r, i, j)

Parameters:

n = number of modularized boxes to be delivered for a given route; w_i = width of modularized box i; h_i = height of modularized box i; d_i = depth of modularized box i;

p_i = visiting order of modularized box i's customer in the visiting sequence;

W = width of a vehicle;

H = height of a vehicle; D = depth of a vehicle;

Decision variables:

x_i : coordinate along the x-axis of the left-bottom-back corner of i;

y_i : coordinate along the y-axis of the left-bottom-back corner of i;

z_i : coordinate along the z-axis of the left-bottom-back corner of i;

l_{ij} : 1 if box i is at the left of box j, 0, otherwise; b_{ij} : 1 if box i is in $b_{ij}$the back of box j, 0, otherwise; u_{ij} : 1 if box i is under box j, 0, otherwise;

u_{ij}: 1 if box I is under box j, 0, otherwise;

Algorithm:

for $1 \leq i, j \leq n$ do

Add the following non-overlapping constraints; $x_i - x_j + W \cdot l_{ij} \leq W - w_i$; $y_i - y_j + H \cdot u_{ij} \leq H - h_i$; $z_i - z_j + D \cdot d_{ij} \leq D - d_i$; if $p_i < p_j$ and $i < j$ then

Add the Last-In-First-Out constraints;

$l_{ij} + l_{ji} + u_{ij} + u_{ji} + b_{ji} = 1$; end if

end for

for $1 \leq i \leq n$ do

Add the following bound constraints;

$W - w_i \geq x_i \geq 0$;

$H - h_i \geq y_i \geq 0$;

$D - d_i \geq z_i \geq 0$;

end for

The second approach is a Bottom-Left-First heuristic algorithm to deal with bin packing.

Algorithm 3 *

Bottom-Left-First like Heuristic to evaluate BinPackingFeasible(r, i, j)

```
Arrange the n modularized boxes based on the increasing order of
pᵢ; I = {0, 0, 0}, L_z = L_x = 0; for i = 1 to n do
    flag = false;
    for (x, y, z) ∈ I do
        if box i can be put at (x, y, z) and x + h_i ≤ L_x, z + d_i ≤ L_z then flag = true,
            break;
        end if
    end for if flag =false
    then
        if L_x = 0 or L_x = H then
            if box i can be put at (0, 0, L_z) then
                x = 0,y = 0,z = L_z,flag =true,L_z = L_z + d_i,L_x = h_i; else
                if L_z < D then
                    L_z = D,L_x = H,i = i - 1;
                end if
            end if
        else
            for (x, y, z) ∈ I : x = L_x, y = 0 do
                if box i can be put at (x, y, z) and z + d_i ≤ L_z then
                    flag =true,L_x = L_x + h_i,break;
                end if
            end for if flag =false
            then
                L_x = H, i = i - 1;
            end if
        end if
    else
        put box i at position (x, y, z), I = I \ {(x, y, z)};
        I = I U {(x + h_i, y, z), (x, y + w_i, z), (x, y, z + d_i)}; end if
end for
```

It is worth noting that in the above Bottom-Left-First heuristic, when the attempt to position box i at point (x,y,z) causes a failure, it is possible that we allow the rotation of the box and try to put the rotated box at (x,y,z) again. Such an extra consideration would increase the chance of *BinPackingFeasible*(r,i,j) being true but would result in a longer computational time as well.

3.2 Numerical Results

As highlighted in the previous section, the main objective of our integrated last mile 3D bin packing problem is to minimize the operational cost while efficiently using the vehicles' capacity and consequently minimizing the number of empty vehicles circulating in the network. To fulfill such a purpose, in the case study, we created two scenarios for comparison. In the first scenario, two logistics service companies serve their individual last mile networks by their own fleet of trucks, while in the second scenario, these two companies pool the truck resource and collaborate to make the last mile delivery with the consolidated customer demand. In this case study, we considered three kinds of modularized boxes (Landschützer et al. 2015). Table 3.1 summarizes the modularized box dimensions used for the case study.

We have created 6 sets of testing instances with the number of customers for the last mile delivery ranging from 10 to 60 (incremental step is 10). In each set, 10 instances are constructed (therefore, there are 60 instances in total). For example, the data in Table 3.2 represent an instance from the set with customer number equal to 10. There are ten arrays separated by square braces (i.e., []), and in each array, the first two elements are the x-y coordinates of a customer's location, and the last three elements of the array represent the total numbers of different types of modularized boxes that the customer demands. For instance, [11,34,16,2,1] stands for that the customer is located at point ($x = 11$,$y = 34$) and the customer requests 16 boxes of type 1 modularized box, 2 type 2, and 1 type 3.

To create an instance for both scenarios, first of all, we randomly generate the locations of a Physical Internet hub and customers who are served by the hub. Then, for each logistics company, the total numbers of boxes for each box type demanded by each customer are also randomly picked up from given ranges. Once the demands for the two logistics companies are generated, for the second scenario, we simply added the corresponding demands and treated the sum as the demands for the horizontal collaboration case. For example, the box demands for the first and second companies are (European Environmental Agency 2018; Montreuil 2010; Campbell and Savelsbergh 2004) and [6,6,0], respectively. Hence, in the second scenario, the box demand for the same customer is [22,8,1].

Table 3.1 Modularized box choices

Box number	Length (m)	Width (m)	Height (m)
1	0.3	0.2	0.2
2	0.3	0.4	0.3
3	0.6	0.4	0.4

Table 3.2 Instance file example

[11,34,16,2,1], [43,−3,6,6,0],
[−41,47,21,1,1], [34,−41,14,3,1],
[−17,9,16,0,0], [−26,−44,18,6,1],
[19,−21,15,4,0], [−17,46,17,1,1],
[−3,33,14,4,0], [5,31,16,5,1]

Table 3.3 Case study results, for ten customers

Instance A	cost A	nVeh B	cost B	nVeh A+B	cost A+B	nVeh AB	cost AB	nVeh
10_1	458	3	445	2	903	5	563	5
10_2	425	2	339	2	764	4	475	4
10_3	338	2	266	2	604	4	433	4
10_4	352	2	345	2	697	4	427	3
10_5	426	2	441	2	867	4	542	4
10_6	437	2	412	2	849	4	621	5
10_7	299	2	356	3	655	5	521	5
10_8	411	2	555	2	966	4	617	4
10_9	380	2	352	2	732	4	545	4
10_10	526	2	470	2	996	4	554	5

Table 3.4 Case study results, for 20 customers

Instance A	cost A	nVeh B	cost B	nVeh A+B	cost A+B	nVeh AB	cost AB	nVeh
20_1	677	4	770	4	1447	8	983	8
20_2	693	4	690	4	1383	8	1000	8
20_3	860	4	850	4	1710	8	1042	8
20_4	627	5	688	5	1315	10	898	9
20_5	686	4	614	4	1300	8	825	8
20_6	750	4	545	3	1295	7	749	7
20_7	760	4	668	4	1428	8	803	8
20_8	671	3	651	4	1322	7	951	8
20_9	769	4	800	5	1569	9	1105	9
20_10	830	4	801	4	1631	8	974	8

First of all, we examine the performances of both approaches to evaluate *BinPackingFeasible*$(r,i,\ j)$. After some trials, it turns out that even for relatively small-scale problems, the constraint programming method spent a few dozen of minutes to get the solutions. In contrast, the Bottom-Left-First heuristic is quite fast. Hence, we only use the Bottom-Left-First heuristic to evaluate *BinPackingFeasible* (r,i,j). Tables 3.3, 3.4, 3.5, 3.6, 3.7, and 3.8 summarize the computational results for the 60 instances. In each table, A_cost and B_cost are the total traveling costs for logistics companies A and B, respectively. A_nVeh and B_nVeh are the numbers of the vehicles that companies A and B need to deploy. A+B_cost is the sum of A_cost and B_cost and A+B_nVeh=A_nVeh+B_nVeh, while AB_cost is the total traveling cost, and AB_nVeh is the total number of vehicles used in the case that logistics companies A and B conduct horizontal collaboration. From this numerical experiment, it can be observed that in terms of total traveling cost, horizontal collaboration is much effective than individual scheduling. On average, compared to the case of individual scheduling, the cost saving rate of horizontal collaboration is 32.3%, which is calculated by the following formula.

Table 3.5 Case study results, for 30 customers

Instance A	cost A	nVeh B	cost B	nVeh A+B	cost A+B	nVeh AB	cost AB	nVeh
30_1	1026	6	1280	5	2306	11	1411	13
30_2	1004	5	1079	5	2083	10	1374	11
30_3	1076	6	948	7	2024	13	1416	12
30_4	885	5	964	6	1849	11	1361	12
30_5	1076	5	951	5	2027	10	1450	11
30_6	1021	6	1062	6	2083	12	1378	12
30_7	932	6	897	6	1829	12	1229	13
30_8	994	6	990	6	1984	12	1407	12
30_9	1273	6	996	6	2269	12	1233	12
30_10	896	5	1004	5	1900	10	1282	10

Table 3.6 Case study results, for 40 customers

Instance A	cost A	nVeh B	cost B	nVeh A+B	cost A+B	nVeh AB	cost AB	nVeh
40_1	1075	7	1199	8	2274	15	1654	16
40_2	1407	8	1360	7	2767	15	1881	15
40_3	1107	8	1335	7	2442	15	1756	16
40_4	1458	8	1222	7	2680	15	1699	16
40_5	1236	8	1207	7	2443	15	1694	16
40_6	1415	8	1262	8	2677	16	1713	16
40_7	1148	7	1177	7	2325	14	1691	15
40_8	1296	8	1244	7	2540	15	1758	15
40_9	1286	7	1293	8	2579	15	1650	15
40_10	1131	7	1160	7	2291	14	1747	14

Table 3.7 Case study results, for 50 customers

Instance A	cost A	nVeh B	cost B	nVeh A+B	cost A+B	nVeh AB	cost AB	nVeh
50_1	1280	10	1436	8	2716	18	2175	18
50_2	1676	9	1830	9	3506	18	2279	18
50_3	1595	8	1650	9	3245	17	1964	19
50_4	1355	9	1797	9	3152	18	2262	19
50_5	1615	9	1481	9	3096	18	2199	18
50_6	1521	9	1468	9	2989	18	2301	19
50_7	1577	8	1602	9	3179	17	2084	18
50_8	1593	9	1757	9	3350	18	2047	18
50_9	1283	8	1444	10	2727	18	2058	19
50_10	1356	8	1456	9	2812	17	1837	17

Table 3.8 Case study results, for 60 customers

Instance A	cost A	nVeh B	cost B	nVeh A+B	cost A+B	nVeh AB	cost AB	nVeh
60_1	1916	11	1951	10	3867	21	2557	23
60_2	1663	10	1882	10	3545	20	2465	22
60_3	1811	11	1896	11	3707	22	2537	24
60_4	1779	11	1813	11	3592	22	2626	22
60_5	1672	11	1965	9	3637	20	2456	22
60_6	1812	11	1870	10	3682	21	2402	23
60_7	1954	11	1925	11	3879	22	2738	22
60_8	1837	10	1768	11	3605	21	2611	22
60_9	2008	12	2418	11	4426	23	2978	24
60_10	2056	11	1889	10	3945	21	2937	23

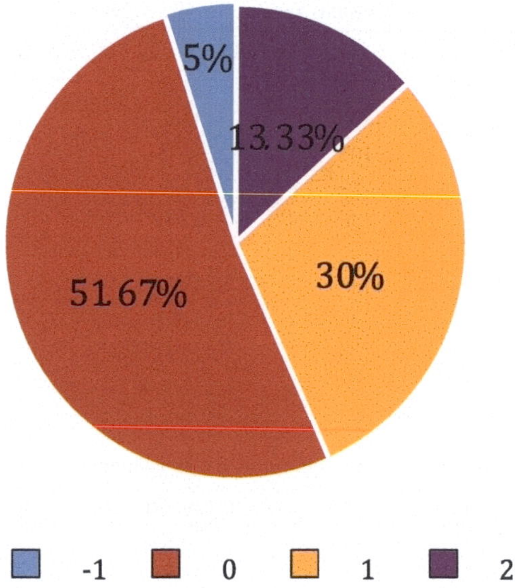

Fig. 3.6 The statistics on the difference of vehicle need on both scenarios

$$\text{cost saving rate} = \frac{\text{A} + \text{B\_cost} - \text{AB\_cost}}{\text{A} + \text{B\_cost}} \times 100\%$$

However, in terms of total number of vehicles used, individual scheduling slightly outperforms collaboration. As depicted in Fig. 3.6, among the 60 testing instances, the percentage that horizontal collaboration uses less vehicles is only 5%. In most of the cases, individual scheduling needs less or equal number of vehicles than horizontal collaboration. However, in around half of the instances, both scenarios ask for the same number of vehicle, and the maximal difference on the number of vehicle used is 2 with percentage 13.33%.

4 Mathematical Model of Vehicle Dispatching Problem

In the following, we will introduce the mathematical optimization model developed for the vehicle dispatching problem in a Physical Internet.
Parameters

- T: an upper bound of the number of arcs for a path for each vehicle.
- K: the set of vehicles.
- $G(N, A)$: the graph representation of a Physical Internet; N is the set of Physical Internet hubs, and A is the set of the arcs connecting hubs.
- $\delta(a) \in N$: the head node of an arc $a \in A$.
- $\sigma(a) \in N$: the tail node of an arc $a \in A$.
- c_a: the traveling cost of arc $a \in A$.
- $l_k \in N$: the initial location of vehicle $k \in K$.
- d_a: the transportation demand for arc a.

Decision Variables

- x_{ka}^{t} : a binary decision variable, 1, if vehicle k takes arc a at tth link, 0, otherwise
- $d_{ka} \in Z^{+}$: the transportation demand portion that vehicle k takes from transportation demand d_a, $a \in A$

Model

$$\min: \sum_{k \in K} \sum_{1 \le t \le T} \sum_{a \in A} c_a x_{ka}^{t} \tag{3.1}$$

s.t.

$$\sum\nolimits_{a \in A} x_{ka}^{t} = 1, \qquad \forall k \in K, 1 \le t \le T \tag{3.2}$$

$$x_{k\acute{a}}^{t+1} \le x_{ka}^{t}, \qquad \forall k \in K, 1 \le t \le T, a, \acute{a} \in A, \sigma(a) \ne \delta(\acute{a}) \tag{3.3}$$

$$\sum_{a \in A : \sigma(a) = l_k} x_{ka}^{1} = 1, \qquad \forall k \in K \tag{3.4}$$

$$\sum\nolimits_{t=1}^{T} x_{ka}^{t} \ge d_{ka}, \qquad \forall k \in K, a \in A \tag{3.5}$$

$$\sum_{k \in K} d_{ka} = d_a \qquad \forall a \in A \tag{3.6}$$

$$x_{ka}^{t} \in \{0, 1\}, d_{ka} \in Z^{+}, \qquad \forall 1 \le t \le T, k \in K, a \in A \tag{3.7}$$

In the above mathematical model, the objective function (3.1) aims to minimize the total traveling distance for all vehicles. Constraints (3.2) make sure that each link of a

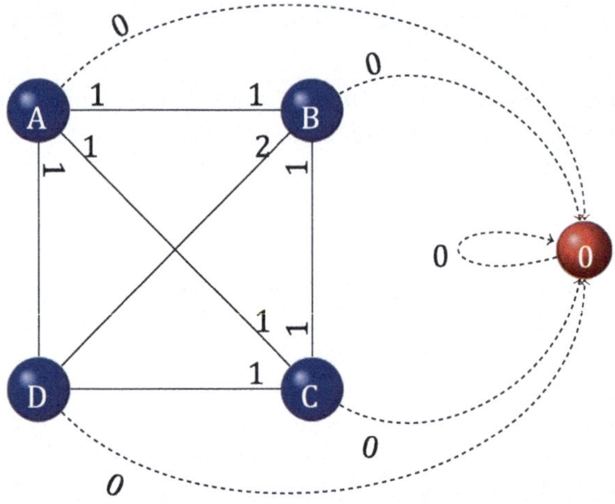

Fig. 3.7 Add dummy node and links for transformation

path for vehicle k should be chosen from the arc set A and only one arc can be chosen. Constraints (3.3) guarantee that a path for vehicle k should be connected.

Specifically speaking, if $x_{ka}^{t} = 1$, that is, vehicle k chooses arc a as the tth link in a path, then the next link in the path should have the starting node same as the ending node of arc a. In other word, if $x_{ka}^{t} = 0$, then $x_{ka}^{t+1}{}_{0}$ should be 0 for all arcs a^{0} with $\sigma(a)$, $\delta(a^{0})$. Constraints (3.4) force that a path of a vehicle k to start from the initial node of the vehicle, i.e., l_k. Constraints (3.5) count the transportation demand fulfilled by a vehicle k. Constraints (3.6) ensure that the total transportation demand for an arc a is split by all vehicles. Finally, constraints (3.7) define the domains for all the decision variables.

However, it should be highlighted that the mathematical model listed above cannot be directly used due to some technical limitations of the selected integer programming modeling framework (i.e., for a vehicle k, x_{ka}^{t} for all $t \leq T$ should be well defined, but T is just an upper bound; therefore, if T is not appropriately chosen, x_{ka}^{t} for all $t \leq T$ cannot be well defined). To bypass such a modeling difficulty, a simple way is to introduce one dummy node 0 and ($|A|$ + 1) dummy arcs with 0 traveling cost and transportation demand to transform the original graph $G(N, A)$ to another associated graph. Figure 3.7 shows the transformed graph of the network given in Fig. 3.1. As illustrated in Fig. 3.7, the dummy arcs (0,0), (A,0), (B,0), (C,0), and (D,0) are included in the new graph. The role of them is to enforce that once a vehicle k select a dummy arcs in {(A,0),(B,0),(C,0),(D,0)}, it cannot choose other real arcs along the path and once the vehicle is "trapped" in the dummy arc set, the only arc it can select is (0,0).

4.1 Numerical Results

To test the developed integer programming model for the vehicle dispatching problem, we used the P&G Switzerland historical sales order data to construct a case study.

As depicted in Fig. 3.8, in this case study, 11 hubs located in Switzerland are used. They are at Frenkendorf, Bremgarten, Ecublens, Frauenfeld, Langenthal, Petit-Lancy, Schmitten, Studen, Sursee, Wangen, and Winznau (nodes A to K, respectively). Figure 3.9 summarizes the information on the transportation demands, the traveling costs for all arcs, and the initial locations of all vehicles. Note that the distances are in the unit of kilometers and there are six vehicles: two at hub B and one at hubs C, E, I, and J.

After network transformation, the developed mathematical model is solved by IBM ILOG CPLEX 12.5 in a Dell M4700 (CPU 2.60 GHz and 8.00 GB RAM), and the minimal cost is 5821 km.

Similar to the last mile problem, for the vehicle dispatching problem, we also want to quantify the benefit of horizontal collaboration. Therefore, we test our cases for two scenarios where there are two distribution companies in our Physical Internet framework. As shown in Fig. 3.10, it can be seen that operator O_1 consists of hubs A, C, G, I, and J and the rest of hubs belongs to operator O_2. Both operators O_1 and O_2 have the same number of vehicles whose initial locations are also indicated in

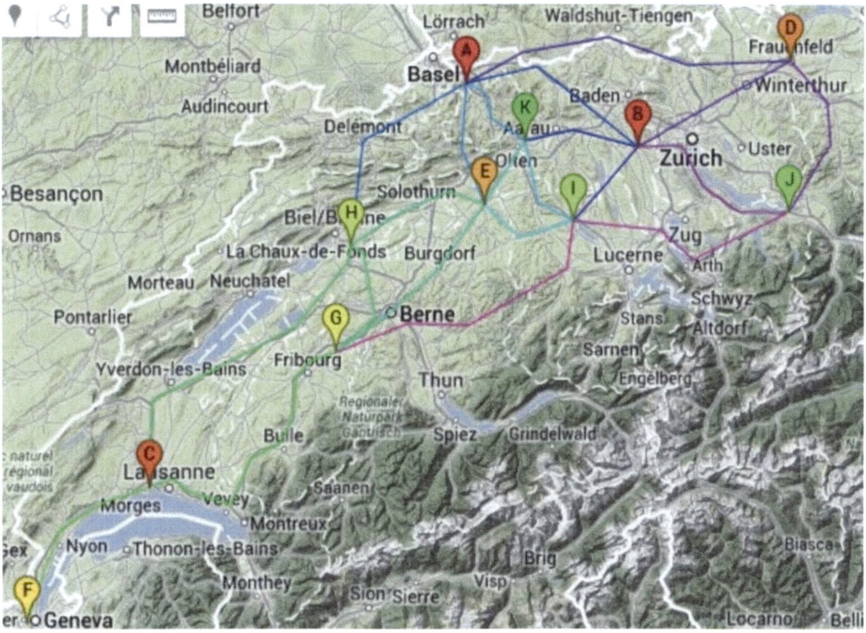

Fig. 3.8 The network of the case study based on P&G Switzerland data

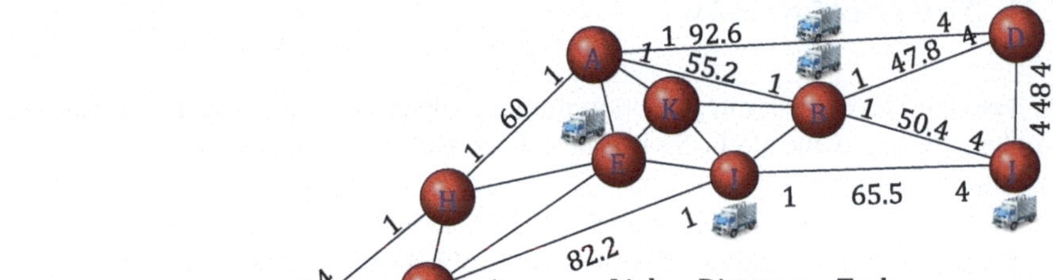

Link	Distance	Task
G-H	35.8	G→H:4,H→G:1
G-E	57.8	G→E:4,E→G:4
H-E	40.8	H→E:1,E→H:4
A-E	33.4	A→E:1,E→A:4
K-E	20.1	K→E:1,E→K:4
K-A	24.0	K→A:1,A→K:1
K-I	27.5	K→I:1,I→K:1
E-I	28.9	E→I:4,I→E:1
K-B	31.9	K→B:1,K→B:1
I-B	26.8	B→I:1,I→B:1

Fig. 3.9 The case study based on P&G Switzerland

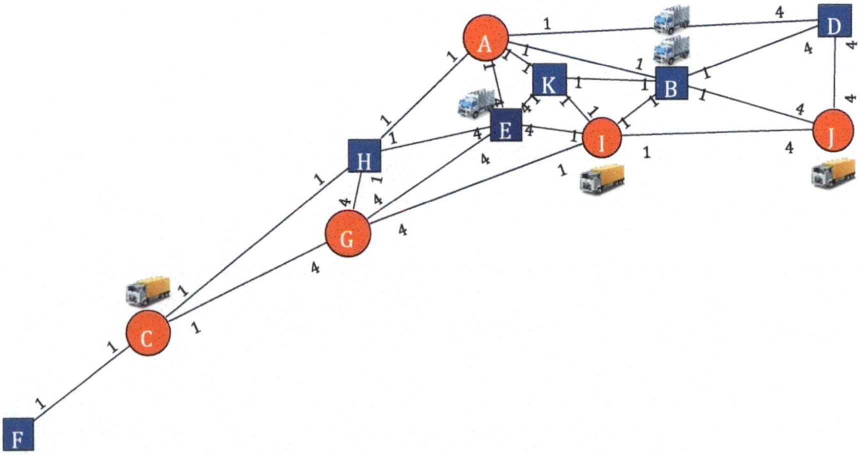

Fig. 3.10 The case study when the network is operated by two operators

Fig. 3.10. Thus, the individual scheduling problems for both O_1 and O_2 are delineated in Figs. 3.11 and 3.12, respectively. After the calculation, we obtain that the optimal costs for both operators are 3822 km and 3387.7 km, that is, 7209.7 km if we took the sum of costs for the two operators. Compared to the scenario of horizontal collaboration, the cost saving amount is 1388.7 km which is equivalent to 19.3% of the total cost for individual scheduling case.

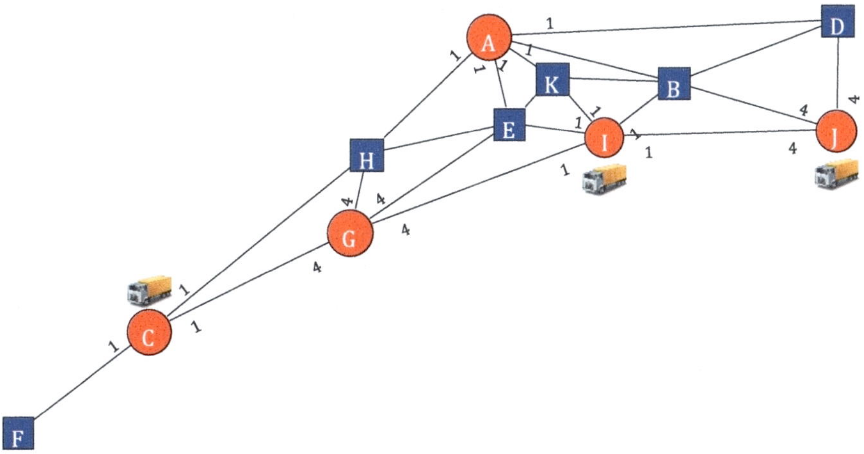

Fig. 3.11 The vehicle dispatching problem for operator O_1

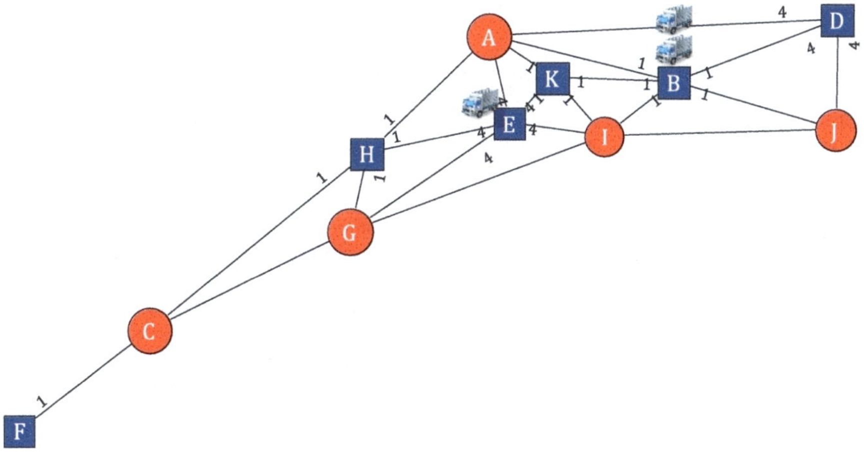

Fig. 3.12 The vehicle dispatching problem for operator O_2

5 Discussions and Conclusion

In this research, we propose two individual optimization problems (i.e., the last mile problem-3D bin packing and the vehicle dispatching problem) in the context of Physical Internet. The last mile problem is a downstream problem which takes the dimensions of the modularized boxes into account for their final distribution to clients. By contrast, the vehicle dispatching problem is a middle stream network problem. Its main objective is to identify the optimal transportation demand split and the best vehicle dispatching plan (i.e., walks for vehicles in the given graph) thus that the cost associated to empty vehicle movements is minimized. Based on these two problems, one of the main purposes of this work has been to quantify the benefit of the horizontal collaboration compared to its counterpart, the individual scheduling, in the world of Physical Internet. We have collaborated with several industrial partners in this project within an EU project. We have been provided with real case studies data which we used to evaluate our algorithms. According to our numerical experiments, the importance and great potential of the horizontal collaboration is highlighted. In the last mile problem, the cost saving rate of the horizontal collaboration can amount to 32%, and in the vehicle dispatching problem, the total vehicle traveling cost can be reduced by 19% if deep collaborations among logistics operators prevail. By reducing the number of empty vehicles circulating in the network, a more sustainable logistic system can be obtained that minimizes environmental impact. We also show the positive effect of horizontal collaborations between different service providers and distributors that make it possible to use resources more efficiently.

References

Alonso-Mora, J., Samaranayake, S., Wallar, A., Frazzoli, E., & Rus, D. (2017). *Proceedings of the National Academy of Sciences, 114*(3), 462.

Bates, O., Friday, A., Allen, J., Cherrett, T., McLeod, F., Bektas, T., Nguyen, T., Piecyk, M., Piotrowska, M., Wise, S., et al. (2018). *Proceedings of the 2018 CHI Conference on Human Factors in Computing Systems* (p. 526). New York: ACM.

Bortfeldt, A. (2012). *Computers & Operations Research, 39*(9), 2248.

Campbell, A. M., & Savelsbergh, M. (2004). *Transportation Science, 38*(3), 369.

European Environmental Agency. (2018). Greenhouse gas emissions from transport. Technical report.

Fanslau, T., & Bortfeldt, A. (2010). *INFORMS Journal on Computing, 22*(2), 222.

Gendreau, M., Iori, M., Laporte, G., & Martello, S. (2006). *Transportation Science, 40*(3), 342.

Gentile, G., & Noekel, K. (2016). Gewerbestrasse: Springer International Publishing.

Hifi, M., Kacem, I., Nègre, S., & Wu, L. (2010). *Electronic Notes in Discrete Mathematics, 36*, 993.

Landschützer, C., Ehrentraut, F., & Jodin, D. (2015). *Logistics Research, 8*(1), 8.

Levine, J., & Ducatelle, F. (2004). *Journal of the Operational Research Society, 55*(7), 705.

Martello, S., Pisinger, D., & Vigo, D. (2000). *Operations Research, 48*(2), 256.

Massen, F., Deville, Y., & Van Hentenryck, P. (2012). *Integration of AI and OR techniques in constraint programming for combinatorial optimization problems* (pp. 260–274). Berlin: Springer.

Montreuil, B. (2010). *Physical internet manifesto V1. 7: Globally transforming the way physical objects are handled, moved, stored, realized, supplied and used*. Québec, CA.

Montreuil, B. (2011). Toward a physical internet: Meeting the global logistics sustainability grand challenge. *Logistics Research, 3*(2–3), 71–87.

Pinedo, M. (2012). *Scheduling: Theory, algorithms, and systems*. New York: Springer.

Zanni, A. M., & Bristow, A. L. (2010). *Energy Policy, 38*(4), 1774.

Chapter 4
Circular Economy in Agri-food Systems

Negin Salimi

Abstract In recent years, demand for food has increased. On the one hand, population, economies, and the urbanization rates have grown increasingly worldwide. On the other hand, increased waste food at all stages of the agri-food systems, including production, storage, processing, wholesale, and consumption, puts enormous pressure on producing as much food as possible. The main results of the inefficiency in meeting the demand for food in agri-food systems are a loss of productivity, energy, and natural resources that finally prove to be expensive. Using the principles of circular economy (CE) provides a kind of solution to minimize waste in agri-food systems and meet demand for food. In other words, implementing CE in agri-food systems leads to a reduction in waste, and there is a possibility to use the waste in question as by-products or co-products in agri-food systems. Applying CE principles in agri-food systems implies implementing policies and decisions at different stages of agri-food systems. In this chapter, we discuss the main decisions and policies that can facilitate the implementation of CE in agri-food systems and help create new opportunities for value-added activities by focusing mainly on the production and consumption of food and on waste management.

Keywords Circular economy · Agri-food · Food production · Food consumption · Waste management

1 Introduction

Agri-food systems comprise different subsectors of crops, livestock, forestry, fisheries, and aquaculture and, using a variety of living organisms and different combinations of land, labor, and capital, convert natural resources into human benefits

---

N. Salimi (✉)
Business Management and Organization, Wageningen University and Research, Wageningen, The Netherlands
e-mail: negin.salimi@wur.nl

J. Rezaei (ed.), *Strategic Decision Making for Sustainable Management of Industrial Networks*, Greening of Industry Networks Studies 8,
https://doi.org/10.1007/978-3-030-55385-2_4

(Campanhola and Pandey 2019). However, agri-food systems have to be adopted themselves based on different contextual changes, such as resource availability, consumer preferences, and the political and intuitional environment (Campanhola and Pandey 2019). More precisely, these contextual changes can ultimately affect the ability of agri-food systems to meet demand for food. Two main influential contextual changes that we focus on in this chapter are food demand and food waste.

It has been predicted that the growing food demand that we are experiencing right now will continue in the future, with the global population expected to add another 2 billion people by 2050. This means that food production also has to grow in order to meet the growth in demand by 2050 (Alexandratos and Bruinsma 2012; Toop et al. 2017). At the same, food waste is one of the main challenges facing global agri-food systems. Food can be wasted at all stages of agri-food systems. More precisely, agri-food systems refer to all the activities and institutions involved in the production, storage, processing, wholesale (distributing), and consumption (Ledger 2016) of food, and at all those stages, food can be wasted in the agri-food systems.

Each year, 700 million tonnes of agricultural waste is generated by Europe alone (Pavwelczyk 2005; Toop et al. 2017). The Food and Agriculture Organization of the United Nations (FAO) (2011) announced that a third of the food that is produced for humans is wasted (FAO 2011). By now, this has become an international issue that has been discussed at many conferences, including the FAO. Moreover, some European institutions, like the European Commission and the European Parliament, have set targets for the reduction of food waste (Halloran et al. 2014). It may be clear that, in addition to boosting the production of food, we also have to be more efficient. If we manage to reduce the amount of food we waste and instead use it as by-products or co-products, that is a big step towards being better able to meet the demand for food.

The question at the moment is how the existing agricultural systems will respond to the abovementioned concerns and challenges. The importance of finding a new approach and new solutions is emphasized by several researchers in this area, for instance, Gunderson (2001) and Berkes et al. (2008).

The circular economy has been introduced in literature as a high-level strategy and as a major opportunity to reduce waste and improving the way resources are used by focusing on creating a closed loop system (Ghisellini et al. 2016; Jurgilevich et al. 2016; Geissdoerfer et al. 2017), in order to improve economic and environmental sustainability (Winkler 2011). The European Commission has defined circular economy as an economy "where the value of products, materials and resources is maintained in the economy for as long as possible, and the generation of waste minimized." Based on this definition, the priority involves circulated items, such as plastics, food waste, critical raw materials, construction materials, biomass, and bio-based products (European Commission 2017).

The circular economy is different from the traditional linear economy, which focuses on "take-make-dispose" (MacArthur 2013). More precisely, in a linear economy, the raw materials and resources that are used in production process are ultimately seen waste, while reducing (minimizing the use of resources), reusing (maximizing the reuse of products), and recycling (recycling the raw materials to a high standard) are three of the main tools in a circular economy designed to create value out of production process as valuable co-products and by-products (Rood and Hanemaaijer 2017).

The circular economy in agri-food systems follows the stream to reuse food and reduce the amount of waste, for example, by using by-products and waste and recycling nutrients (de Boer and van Ittersum 2018). CE has some principles that prevent having waste and surplus out of process, and, if the system contains waste and surplus, some policies and technologies should be applied to minimize them as much as possible. Toop et al. (2017) argue that it is necessary to have a comprehensive system at the food supply chain level and apply CE to agri-food systems as a whole, rather than at the level of individual companies, which requires an evolution in existing production systems and technologies.

The aim of these evolutionary policies is to help the food systems to use more regenerative resources, such as wind and solar energy, instead of using more limited resources such as phosphate rock and land (Van Zanten et al. 2019). Additionally, to move towards CE in agri-food systems, we have to look for policies that prevent leaking natural resources like nitrogen and phosphorus from the system and ultimately adding value to the system by reusing or recycling resources like human excreta (Van Zanten et al. 2019). It is important to consider the way the decisions made by various actors affect one another in the agri-food systems in terms of the circular economy. More interestingly, their decisions are dependent on each other, which means that a decision that is made by one actor in the food systems affects the decisions made by others and a decision that is made in line with sustainability by one actor can help other actors become more sustainable and vice versa (Halloran et al. 2014). This can satisfy one of the requirements of applying CE in agri-food systems, which is making all decisions in the agri-food supply chain more sustainable.

Before discussing the main principles of CE and policies that make it possible to apply its principles to agri-food systems, we highlight some of the characteristics of agri-food systems that it is important to consider before applying CE.

2 Key Characteristics of Agri-food Systems

Thompson et al. (2007) identified the following characteristics of agri-food systems:

1. The dynamics of production: this means that agriculture is dependent on different variables like natural resources, spatial dispersion of activity, seasonal variables, and information asymmetries due to location and distance.
2. Integrated agri-food systems: this is linked to agri-food systems, as regional and global supply chains, connecting producers and consumers from different parts of the world, which means that agri-food systems in one part of the world are affected by technological changes and developments, political decisions, and institutional changes elsewhere.

3. Market failures: failures in the market for agricultural products can be the result of different factors, like high transaction cost and information asymmetries.
4. Public sector interventions: it is necessary to compensate the market failure agri-food systems by public sector interventions.
5. Sociocultural systems: agriculture is a combination of "agri," which is related to a productive activity, and "culture," which is connected to a way of life. This means that social relations are essential to accessing resources, power asymmetries, and benefits from public services.
6. Heterogeneity and diversity: in agriculture, there is diversity among actors. Some actors have access to better assets, to better technologies, and, ultimately, to better new markets, while many other actors have access to limited resources and a limited number of consumers.
7. Collective action: different types of collective and cooperative actions are important for farmers to have their voices reach the policy-makers, including their request for accessing public resources, managing common property resources, and meeting new market requirements.

In the following section, we discuss some of the main challenges that exist in agri-food systems. CE can be a solution to handling these challenges and arrive at more sustainable agri-food systems.

3 Challenges for Agri-food Systems

Jurgilevich et al. (2016) mentioned the main problems and challenges in three stages of agri-food systems by considering the circular economy: food production, food consumption, and food waste and surplus management.

The first stage in agri-food systems, based on the model proposed by Jurgilevich et al. (2016), is food production, for which we need fertilizer containing nutrients like nitrogen, phosphors, and potassium (Cordell et al. 2009). However, due to a growing population and a change in the human diet from plant-based to meat-based, demand for nutrients, specially phosphors, has increased (Elser 2012), which poses an important challenge to sustainable phosphors management (Metson et al. 2012). More precisely, phosphate rocks are the main source of phosphor, which is a nonrenewable resource, and it is predicted that in the next 50–100 years, existing global reserves will be used up (Cordell et al. 2009). Additionally, changes in agricultural methods to increase production by intensification of the use of fertilizer increases the scarcity of phosphor resources (Godfray et al. 2010). As such, it is essential to look for policies, methods, and technologies to recover and reuse phosphors from different resources, like human and animal excreta and food and crop waste (Jurgilevich et al. 2016; de Boer and van Ittersum 2018).

Food consumption is the next stage in the agri-food systems, based on Jurgilevich et al. (2016), that creates problems in terms of applying the circular economy. One of the most challenging problems in food systems is the growing rate of meat consumption among humans. According to the United Nations report, meat production consumes high levels of natural resources and energy (Motavalli 2008; Jurgilevich et al. 2016), in addition to producing greenhouse gases and other pollutants in greater quantities than the transportation system (Motavalli 2008). Moreover, Jurgilevich et al. (2016) specify that not only do consumers not have access to information about the actors involved in the food production system, they also are not aware of the ethical consequences of their food-related choices. Some of these problems are related to different factors, like tradition and culture, and value and belief systems.

The final stage of agri-food systems is that of food waste and surplus management. Food surplus is defined as extra but edible food (Papargyropoulou et al. 2014). When this edible food becomes inedible or is lost, the result is waste food. In other words, a waste of food is a waste of natural resources (e.g., water and land) and agricultural input (e.g., the fertilizers, energy, and labor that are used in the different stages of food systems, like processing, distribution, storage, and consumption) (Halloran et al. 2014) and even packaging (Parfitt et al. 2010). "Packaging design for sustainability" has been considered one of the important issues in the field of sustainability (Duizer et al. 2009). Rezaei et al. (2019) argue that, to have a sustainable packaging design, it is more important to take the criteria into account that guarantee the sustainability of the entire supply chain than merely looking at product- or company-specific sustainability criteria.

FAO (2011) reports that, in industrialized countries, much more food waste is created on a per capita basis than in developing countries. Additionally, based on this report, in high-income countries, waste occurs more at the consumption stage, while in low-income countries, more waste is created in the primary stages of food systems.

Some agricultural waste may be avoidable (can be prevented), while other waste is unavoidable (cannot be prevented). Knowing these features can help improve the strategy and plan in advance to prevent food waste, especially when it is avoidable, or reuse/recycle unavoidable waste. In the AgroCycle project (AgroCycle 2017), agricultural waste, co-products, and by-products (AWCB) are divided into avoidable and unavoidable (see Table 4.1).

It is not only some direct actions, like a lack of coordination and knowledge being shared among actors in the agri-food systems that cause food waste (FAO 2011), but also the indirect actions of some of the actors in agri-food systems, for instance, through packaging sizes, sale promotions, or discounts that affect the behavior of consumers (Parfitt et al. 2010).

Figure 4.1 shows different types of waste that can be produced at different stages of food systems in Denmark (Halloran et al. 2014), which we think to a large extent applies to other parts of the world as well.

Based on Fig. 4.1, the producers of basic food products, from large-scale commercial producers to family farmers, are involved and active in the primary sector. At this stage, the most significant loss of edible products takes place, including dead animals and wasted seeds in the field. The major types of food waste in the food processors stage are related to dairy products and meat, with only a minor share of

Table 4.1 Different types of agricultural wastes, co-products, and by-products (AWCB) (AgroCycle 2017)

AWCB type	Description
Avoidable	Avoidable AWCB are material streams that have been mismanaged and disposed of and are typically a mixture of different components (heterogeneous). These include wasted foods generated in processing, retail, catering, and households. Avoidable "wasted food" occurs when foods are discarded because they are regarded as "suboptimal," or when they pass their "best before" date, or due to product flaws
Unavoidable	Unavoidable AWCB, on the other hand, are materials arising from food production systems that are not consumable, typically described as by-products, co-products, or residues (e.g., manures, crop residues, leaves, and peels). Unavoidable AWCB cannot be prevented and are typically homogeneous streams

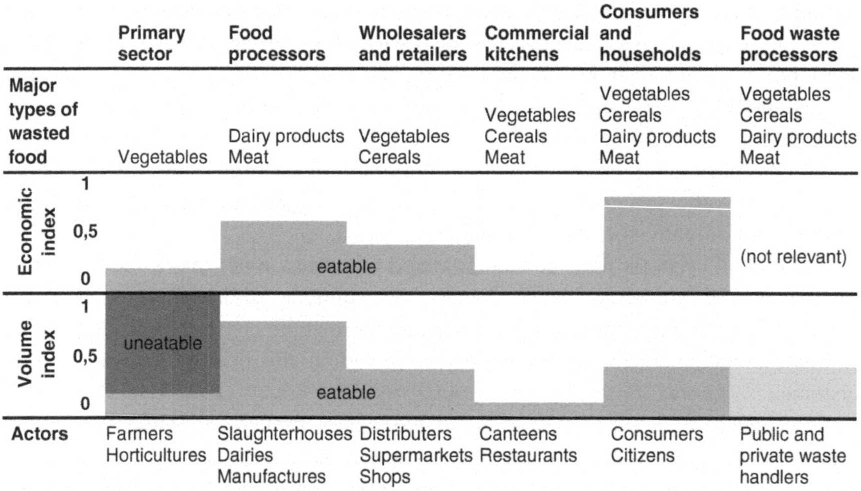

Fig. 4.1 Different types of food products wasted at different stages in food systems in Denmark. (Halloran et al. 2014)

the waste being edible (only 1–2% of overall production), while the rest of the food waste is inedible and viewed as by-products. Other actors producing waste are wholesalers and retailers, including warehouses and supermarkets. At this stage, most of the waste products are bread, yoghurt, vegetables, and fruit. Kitchens where large quantities of foods are being prepared for large groups of people, like the kitchens in hotels, universities, and hospitals, are another source of food waste, while consumers and households also play a significant role. At this stage, the food being wasted can be either edible or inedible. Finally, waste can come from public and private waste handlers. In fact, retailers and wholesalers have an agreement with private companies to transport and incinerate the waste, with the aim of energy recovery. However, there are different types of wasted food, from vegetables to meat products, involved in that process (Halloran et al. 2014).

4 Applying Circular Economy in the Food Systems

The idea of circularity originated in industrial ecology, which aims to reuse or recycle materials and, in doing so, reduce resource consumption (Jurgilevich et al. 2016). To apply circular economy in food systems, we need certain decisions and policies at different stages of food systems to prevent food being wasted, while, in cases where such loss is inevitable, we need policies designed to recover, reuse, remanufacture, or recycle the food material in question. Shove (2014) notes that, for such a sustainable evolution, it is essential to change technologies and infrastructures to maximize the use of regenerative resources like wind and solar energy, instead of using limited resources like phosphate rocks and land. Additionally, the loss of resources like nitrogen and phosphorus from food systems should be prevented using technologies and policies (Van Zanten et al. 2019). Finally these policies should motivate the actors involved to reuse or recycle resources like human excreta (Van Zanten et al. 2019). It is not only essential to consider changes in technologies to move towards a circular direction in food systems, we also need to use different strategies, including education, to persuade motivate consumers to choose a plant-based diet rather than a meat-based diet (Jurgilevich et al. 2016).

In the following subsections, we discus some of the existing potential policies, practices, and decisions in relation to circular food systems.

4.1 AgroCycle for a Circular Economy

AgroCycle is a Horizon 2020 project that aims to deliver a sustainable value chain by using agricultural waste, co-products, and by-products (AWCB) in an efficient manner. The AgroCycle project includes 25 partners from the EU, Hong Kong, and China who conduct research in different scientific areas: fruit, animal and dairy science, horticulture, recycling, sustainable economy, and dairy (AgroCycle 2017).

This project is based on an integral analysis of the entire agri-food value chain by looking at livestock and crop production, food processing, and the retail sector (see Fig. 4.2). In fact, the project's main goal is to use mechanisms to valorize agricultural waste for other uses (AgroCycle 2017).

Using and extracting animal manures and food waste from agriculture are not something new. Some of the practices, like composting and anaerobic digestion, have already been used for many years in traditional valorization technologies. In addition, there are many traditional and innovative valorization pathways to exploit the AWCB being produced each year by agriculture, food processing, and domestic waste generation. However, increasing the value of agri-food chain requires a holistic approach and framework. To reach that goal, we need to examine the importance of using AWCB in agricultural systems and the AWCB affect the environment. More importantly, it is essential to conduct analyses and studies to determine where it is possible to remove AWCB from the chain without having any

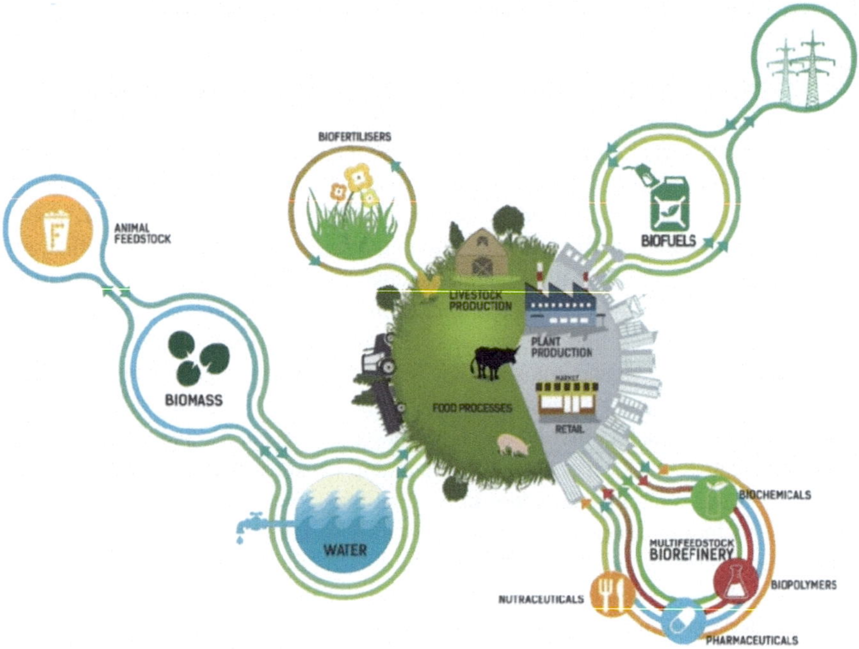

Fig. 4.2 AgroCycle innovations in the agricultural production chain. (Toop et al. 2017)

negative effect on the performance of agricultural system. And finally, it is essential to examine which AWCB are right and available for valorization (AgroCycle 2017).

4.2 Optimal Use of All Biomass

Van Zanten et al. (2019) specify that the use farm animals contributes to circular food systems. They introduced the biophysical concepts of circularity in the food systems and determined that the basic building block of circular food systems is plant biomass (see Fig. 4.3).

Based on Fig. 4.3, arable land is used for the production of nutritious food from plant biomass. Biomass that is not fit for human consumption is used as animal feed through recycling. To provide soil fertility, by-products and manures, like food waste, animal and human excreta, and crop residues, are used. In fact, by-products are produced during the production and consumption of foods from plant material (Van Zanten et al. 2019). As mentioned earlier, these by-products are either edible or inedible for humans (Halloran et al. 2014). To apply circularity in food systems, we need to prevent by-products that humans can consume by reusing them as food for people. Finally, unavoidable edible and inedible by-products should be recycled to fertilize soil or feed animals. Through this process, not only can we use the

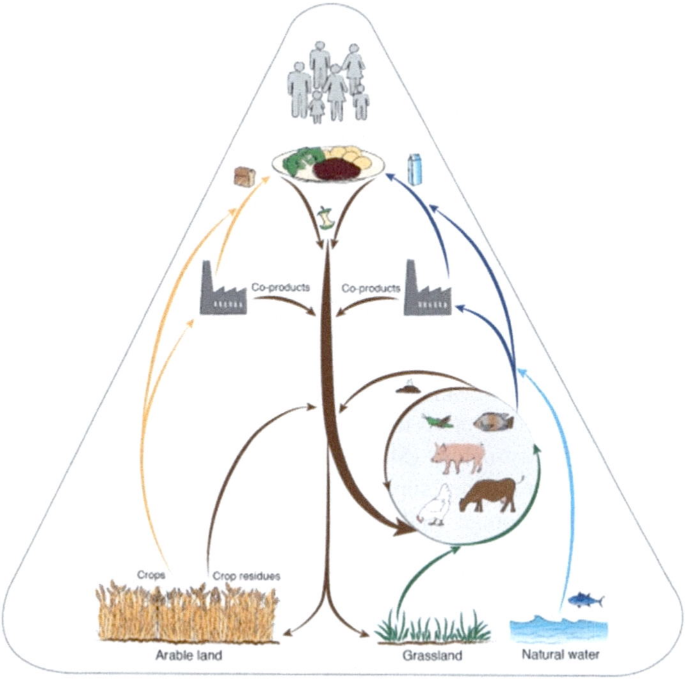

Fig. 4.3 The biophysical concept of circularity. (Van Zanten et al. 2019)

contribution of animals to a circular food system and recycling nutrients, it is also possible to feed the future population through sustainable food systems (Van Zanten et al. 2019).

4.3 Local Food Systems

Local food systems are a strategy designed to improve the quality of people lives (Allen 2010). The concept of local food systems contrasts with global food systems, where the consumers and producers of food do not have a direct relationship and act at a distance (Hinrichs 2000). Direct connections between producers (farmers) and consumers in local food systems promote food quality, equity, and security. Additionally, local food systems provide a way to develop sustainable food systems. More precisely, by applying a local food strategy, we can reduce the waste, recycle nutrients, and reduce storage and transportation cost by shortening the supply chain. Additionally, through local food systems, food security can be enhanced via specialization in regional products (Jurgilevich et al. 2016).

Hinrichs (2000) discusses two types of direct agricultural markets: farmer's markets and community-supported agriculture. The former category is characterized by tight social relations based on familiarity, habit, and sentiment between farmers and consumers, and these relations remain at the commodity level alone. Community-supported agriculture, on the other hand, is based on shared partnership and agreement among local farmers and consumers to share costs, risks, and products. Groh and McFadden (1998) indicates that the main difference between community-supported agriculture and other types of direct markets is that community-supported agriculture focuses on making community-related food, land, and nature. White and Stirling (2013) indicate that community-supported agriculture is "a critical arena for exploring trajectories towards Sustainability."

There are some barriers for farmers and producers of local food to satisfy the demand from all consumers. Martinez (2010) explained that the producers of local food will not always manage to do so, either due to limited capacity or due to demand for food that is unavailable or out of season, high demand for specific products, or the fact that delivery is time-consuming. One another barrier is a lack of proper distribution infrastructure (Shipman 2009). In fact the farmers and food producer do not access the supply chain infrastructure, like vehicles or temperature-controlled storage facilities (Martinez 2010). Because, economically speaking, it make more sense for local farmers to combine their products and foods with the products of other local farmers at the processing and shipping stage, it is difficult to have a traceback mechanism in place (Martinez 2010). However, product traceability is very important to buyers, because it provides them with information about the origin of products and the risk of food-related illnesses (Shipman 2009). The final barrier for local farmers is their limited expertise and training concerning important and relevant topics like risk management, food safety liability, and good agricultural practices. Training in these areas is of special importance in terms of the potential success, especially for young farmers (Shipman 2009).

4.4 Waste Management

A number of countries in the world now focus on reducing the amount of food being wasted. Since 2011, that has also been the aim of a number of European Institutions (EP 2011). In the following section, we list some of solutions and regulations that can be used to reduce food waste at different stages of agri-food systems:

- Providing enough information and incentives helps retailers predict the amounts of food they need to order from producers and wholesalers (Nordic Council of Ministers 2011).
- The use of a "just-in-time" approach by primary agricultural producers, especially for products like milk and fruits, which need to be fresh and delivered in time. This method provides greater certainty when it comes to ordering the right amount of products (Halloran et al. 2014).

- To apply a "just-in-time" strategy, there are some requirements and conditions, for instance, coordination between production and logistics and sharing knowledge between the two can help minimize avoid waste (Halloran et al. 2014).
- Creating and increasing trust and loyalty between retailers and producers will help reduce food waste (Weber et al. 2011).
- Understanding the buying habits and behavior of consumers helps reduce food waste. Essentially, consumers decide to buy products based visual aspects, which means that design and packaging are important factors. Also, offering the right packaging size may increase the consumption of a product before it goes to waste (Halloran et al. 2014).

5 Conclusions

In this chapter, we discussed a very challenging issue in the world: circular economy in agri-food systems. We discussed the main problems involved in applying a circular economy in agri-food systems, in light of the specific characteristics of this system, and examined what regulations can help create agri-food systems where materials and products maintain their value as much as possible.

Applying a circular economy to agri-food systems in which different actors work together in different ways and where their decisions affect one another means we need to start by looking at the actors involved and motivate them to use finite and shrinking natural resources in ways that are sustainable (Halloran et al. 2014). It is essential for the different actors in our food systems to work together and share knowledge (FAO 2011). For instance, the collaboration and coordination among the actors involved in production and those involved in logistics are needed to reduce and prevent waste as much as possible. Data and knowledge directly affect the amount of waste being produced, and it is important for the actors to come together and work out how and what data to share (Halloran et al. 2014).

At the food production stage, it is important to explore the recovery of phosphors. Schösler et al. (2015) suggested using new policies and technologies to compensate all exported phosphor from farms through waste instead of using phosphate rock. Moreover, localized food systems improve the quality, equity, and security of food. As such, policy regulations, like implementing tax incentives for recovering nutrients and encouraging local farmers to sell their products locally, can bring us closer to a circular economy in agriculture (Jurgilevich et al. 2016). In fact, tax incentives may provide farmers with the funds they need to operate their business locally.

At the food consumption stage, consumers play a crucial role in the choices they make. In fact, moving people away from a meat-based diet (which uses huge amounts of nutrients and energy) to a plant-based one (which is more sustainable) will make food systems more sustainable. Jurgilevich et al. (2016) suggest providing educational programs to consumers to make them aware of the consequences of their food-related choices and show them how they can control and prevent waste at this stage. Moreover, it is important for producers to be transparent towards consumers and share information about the sustainability of production processes and manufacturers in the form of product labels, allowing consumers to make a more informed and sustainable decision. De Boer et al. (2014) also indicate that this goal (having closed-loop food systems) can be realized with the help of supermarkets, food services, and public food catering, who can make plant-based foods more attractive to consumers through advertisements, promotions, or discount plans.

Finally, to implement the concept of a circular economy, we need a more holistic approach in which we consider sustainability at all the different stages of food systems, from farming to consuming, where the importance of each actor's role and decisions cannot be ignored by the other actors. Regardless of the problems and challenges involved in applying a circular economy to agri-food systems, it is important for the actors involved in food systems to work together on the basis of trust, commitment, and knowledge sharing.

References

Agrocycle. (2017). *Agrocycle 2017* [Online]. Available: www.agrocycle.eu. Accessed 7 Jan 2020.

Alexandratos, N., & Bruinsma, J. (2012). *World agriculture towards 2030/2050: the 2012 revision.*

Allen, P. (2010). Realizing justice in local food systems. *Cambridge Journal of Regions, Economy and Society, 3*(2), 295–308.

Berkes, F., Colding, J., & Folke, C. e. (2008). *Navigating social-ecological systems: Building resilience for complexity and change.* Cambridge: Cambridge University Press.

Campanhola, C., & Pandey, S. (2019). *Sustainable food and agriculture* (pp. 305–330). London: Academic.

Cordell, D., Drangert, J. O., & White, S. (2009). The story of phosphorus: global food security and food for thought. *Global environmental change, 19*(2), 292–305.

de Boer, I. J., & van Ittersum, M. K. (2018). Circularity in agricultural production. In *Animal production systems and plant production systems.* Wageningen: Wageningen University and Research.

de Boer, J., Schösler, H., & Aiking, H. (2014). "Meatless days" or "less but better"? Exploring strategies to adapt Western meat consumption to health and sustainability challenges. *Appetite, 76*, 120–128.

Duizer, L. M., Robertson, T., & Han, J. (2009). Requirements for packaging from an ageing consumer's perspective. *Packaging Technology and Science: An International Journal, 22*(4), 187–197.

Elser, J. J. (2012). Phosphorus: a limiting nutrient for humanity? *Current Opinion in Biotechnology, 23*(6), 833–838.

EP. (2011). *How to avoid food wastage: strategies for a more efficient food chain in the EU.* Brussels: European Parliament.

European Commission. (2017). Report from the Commission to the European Parliament, the Council, the European Economic and Social Committee and the Committee of the Regions on the Implementation of Circular Economy Action Plan, Brussel, 26.1.2017, Com (2017) 33 final.

FAO. (2011). *Global food losses and food waste—Extent, causes and prevention.* Rome: FAO.

Geissdoerfer, M., Savaget, P., Bocken, N. M. P., & Hultink, E. J. (2017, February 1). The circular economy – A new sustainability paradigm? *Journal of Cleaner Production, 143*, 757–768. https://doi.org/10.1016/j.jclepro.2016.12.048.

Ghisellini, P., Cialani, C., & Ulgiati, S. (2016). A review on circular economy: the expected transition to a balanced interplay of environmental and economic systems. *Journal of Cleaner Production, 114*, 11–32.
Godfray, H. C. J., Beddington, J. R., Crute, I. R., Haddad, L., Lawrence, D., Muir, J. F., Pretty, J., Robinson, S., Thomas, S. M., & Toulmin, C. (2010). Food security: The challenge of feeding 9 billion people. *Science, 327*(5967), 812–818.
Groh, T., & McFadden, S. (1998). *Farms of tomorrow revisited*. Great Barrington: SteinerBooks.
Gunderson, L. H. (2001). *Panarchy: Understanding transformations in human and natural systems*. Washington: Island Press.
Halloran, A., Clement, J., Kornum, N., Bucatariu, C., & Magid, J. (2014). Addressing food waste reduction in Denmark. *Food Policy, 49*, 294–301.
Hinrichs, C. C. (2000). Embeddedness and local food systems: notes on two types of direct agricultural market. *Journal of Rural Studies, 16*(3), 295–303.
Jurgilevich, A., Birge, T., Kentala-Lehtonen, J., Korhonen-Kurki, K., Pietikäinen, J., Saikku, L., & Schösler, H. (2016). Transition towards circular economy in the food systems. *Sustainability, 8* (1), 69.
Ledger, T. (2016). Power and Governance in Agri-Food Systems: Key Issues for Policymakers. *Trade & Industrial Policy Strategies, 23*.
MacArthur, E. (2013). *Towards the circular economy, economic and business rationale for an accelerated transition*. Cowes: Ellen MacArthur Foundation.
Martinez, S. W. (2010). *Local food systems; concepts, impacts, and issues (No. 97)*. Diane Publishing.
Metson, G. S., Bennett, E. M., & Elser, J. J. (2012). The role of diet in phosphorus demand. *Environmental Research Letters, 7*(4), 044043.
Motavalli, J., 2008. The meat of the matter. *E-NORWALK-, 19*(4), p. 26.
Nordic Council of Ministers. (2011). *Initiatives on prevention of food waste in the retail and wholesale trades*. Copenhagen: Nordic Council of Ministers.
Papargyropoulou, E., Lozano, R., Steinberger, J. K., Wright, N., & bin Ujang, Z. (2014). The food waste hierarchy as a framework for the management of food surplus and food waste. *Journal of Cleaner Production, 76*, 106–115.
Parfitt, J., Barthel, M., & Macnaughton, S. (2010). Food waste within food supply chains: Quantification and potential for change to 2050. *Philosophical Transactions of the Royal Society B: Biological Sciences, 365*(1554), 3065–3081.
Pavwelczyk, A. (2005). EU Policy and Legislation on recycling of organic wastes to agriculture. *International Society for Animal Hygiene, 1*.
Rezaei, J., Papakonstantinou, A., Tavasszy, L., Pesch, U., & Kana, A. (2019). Sustainable product-package design in a food supply chain: A multi-criteria life cycle approach. *Packaging Technology and Science, 32*(2), 85–101.
Rood, T., & Hanemaaijer, A. (2017). *Opportunities for a circular economy*. The Hague: PBL Netherlands Environmental Assessment Agency.
Schösler, H., de Boer, J., Boersema, J. J., & Aiking, H. (2015). Meat and masculinity among young Chinese, Turkish and Dutch adults in the Netherlands. *Appetite, 89*, 152–159.
Shipman, D. (2009). *Setting the stage: Local foods issues and policies, presentation at local food systems: Emerging research and policy issues conference at USDA*. Washington, DC: Economic Research Service.
Shove, E. (2014). Putting practice into policy: reconfiguring questions of consumption and climate change. *Contemporary Social Science, 9*(4), 415–429.
Thompson, J., Millstone, E., Scoones, I., Ely, A., Marshall, F., Shah, E., Stagl, S., & Wilkinson, J. (2007). *Agri-food systems dynamics: Pathways to sustainability in an era of uncertainty*.
Toop, T. A., Ward, S., Oldfield, T., Hull, M., Kirby, M. E., & Theodorou, M. K. (2017). AgroCycle – Developing a circular economy in agriculture. *Energy Procedia, 123*, 76–80.
Van Zanten, H. H., Van Ittersum, M. K., & De Boer, I. J. (2019). The role of farm animals in a circular food systems. *Global Food Security, 21*, 18–22.

Weber, L., Mayer, K. J., & Macher, J. T. (2011). An analysis of extendibility and early termination provisions: The importance of framing duration safeguards. *Academy of Management Journal, 54*(1), 182–202.
White, R., & Stirling, A. (2013). Sustaining trajectories towards Sustainability: Dynamics and diversity in UK communal growing activities. *Global Environmental Change, 23*(5), 838–846.
Winkler, H. (2011). Closed-loop production systems—A sustainable supply chain approach. *CIRP Journal of Manufacturing Science and Technology, 4*(3), 243–246.

Chapter 5
Location Selection of Bioethanol Distribution Centers

Siamak Kheybari and Alireza Pooya

Abstract In this chapter, the location problem of bioethanol distribution centers is investigated based on two objectives. While the first objective is to maximize the overall utility of the selected places, the second objective is to minimize the cost allocated to establishing the distribution centers. To solve the problem, a hybrid methodology consisting of the best-worst method (BWM) and a set covering model is proposed. BWM is employed to calculate the overall utility of candidate places, and the set covering model is used to select the minimum number of places. By employing the methodology, the selected distribution centers cover the demand of customers within a specific radius. To calculate the overall utility of the candidate places using BWM, a comprehensive framework of sustainability criteria is presented. The proposed framework is formed by criteria extracted from studies which have employed multi-criteria decision-making methods to solve the location problem of distribution centers in different industries. The result of BWM in the proposed methodology is employed as a parameter in the utility function of the set covering model. To solve the model, Lp-metric is suggested. The performance of the proposed methodology is evaluated by a set of data collected from Iran. The proposed hybrid methodology can also be used for locating distribution centers in other industries.

Keyword Bioethanol · Distribution centers · Best-worst method (BWM) · Set covering model · Lp-metric

---

S. Kheybari (✉)
Department of Management, Ferdowsi University of Mashhad, Mashhad, Iran

NEOMA Business School, Mont-Saint-Aignan Cedex, France
e-mail: Siamak.Kheybari@neoma-bs.fr

A. Pooya
Department of Management, Ferdowsi University of Mashhad, Mashhad, Iran

J. Rezaei (ed.), *Strategic Decision Making for Sustainable Management of Industrial Networks*, Greening of Industry Networks Studies 8,
https://doi.org/10.1007/978-3-030-55385-2_5

1 Introduction

Biofuel produced from urban and industrial waste, crops, and remains of forests is generally categorized into three groups including gas, liquid, and solid (Ben-Iwo et al. 2016). Coping with air pollution which usually originates from fossil fuels is one of the most important features of biofuels (Prasad et al. 2012). From among the three categories, the demand for liquid biofuels due to high flexibility in consumption and also the ability to use in transportation sector is significantly increasing in today's world. Ethanol is employed in a variety of industries such as chemical, pharmaceuticals, and cosmetics (Saini et al. 2015). The application of bioethanol in different industries is briefly presented in Fig. 5.1.

For easy supply of bioethanol, determining the optimal location of its distribution centers is an important decision. In such a location problem both (i) utility of candidate places defined by criteria categorized into economic, social, and environmental dimensions (Santibañez-Aguilar et al. 2014) and (ii) coverage radius of the selected centers, which is a requirement to meet customers' needs at the right time

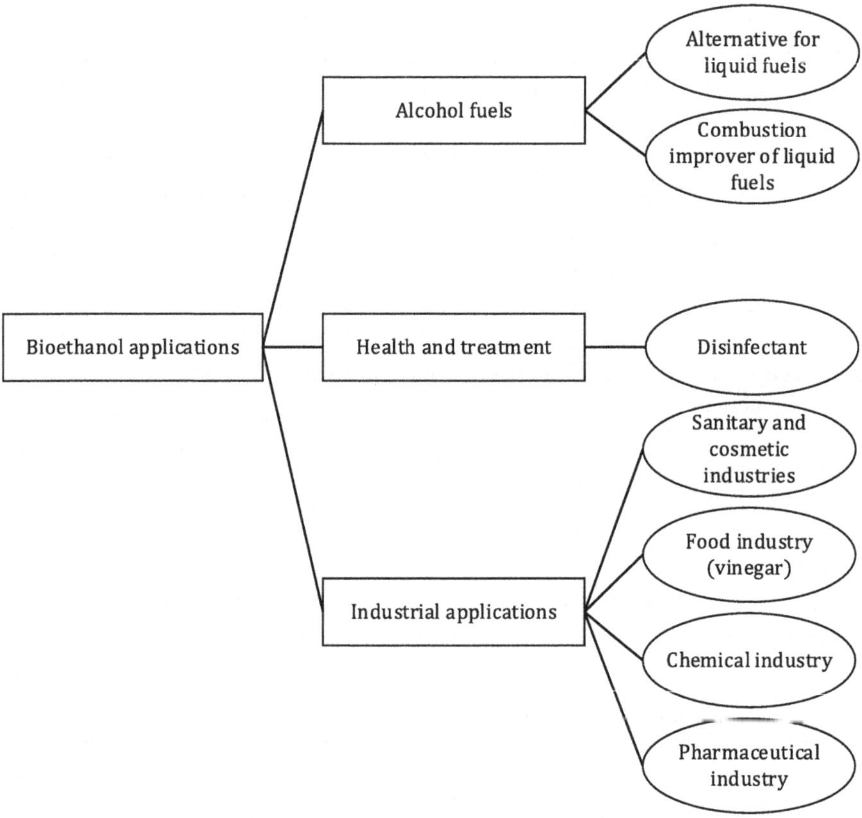

Fig. 5.1 Applications of bioethanol (Yoruklu et al. 2019)

(Kemp 2008), should be simultaneously considered with the lowest cost. To meet the first objective, a comprehensive framework of criteria which are used for this decision needs to be formed. Such a framework increases the accuracy of the selected locations (Bhushan and Rai 2007). However, although in existing literature there are many criteria under social, economic, and environmental dimensions of sustainability to calculate the utility of distribution centers, there is no a comprehensive framework which covers all the dimensions. To this end, a comprehensive framework of sustainability criteria which is the extended version of the framework presented by Kheybari et al. (2019a) is used in this study. By using the proposed framework and a hybrid multi-criteria decision-making (MCDM) method, the utility of candidate places is calculated.

To satisfy the second objective discussed, a hybrid methodology including the best-worst method (BWM) (Rezaei 2015) and a set covering model is presented. Through the proposed methodology, minimum number of places with the maximum utility can be selected so that customers' need is satisfied at the right time.

The rest of this chapter is organized as follows: In Sect. 5.2, the literature is reviewed and the sustainability framework is presented. In Sect. 5.3, the suggested methodology is discussed. In Sect. 5.4, the proposed methodology is employed to determine the appropriate location of bioethanol distribution centers in Iran, and finally the conclusion is presented in Sect. 5.5.

2 Literature Review

In this section, the literature is reviewed in order to identify criteria which have effect on the utility of candidate places. In this regard, papers employing MCDM methods to select distribution centers in different industries were reviewed. The text and tables of the reviewed papers were used to identify criteria which had effect on the location of bioethanol distribution centers. As the result of the literature review, a comprehensive framework is presented in Table 5.1. To categorize criteria into economic, social, and environmental dimensions, both references and criteria's definition were applied. The papers reviewed are discussed as follows.

Chen (2001) used a fuzzy set theory and fuzzy preferences approach in order to investigate factors affecting the location of distribution centers in logistic companies. Alternatives in that research were evaluated by five criteria including *investment cost, expansion possibility, closeness to demand market, availability of acquirement material*, and *human resource*. Chan and Chong (2004) suggested analytic hierarchy process (AHP) and genetic algorithm (GA) to select the appropriate location of distribution centers in a supply chain. *Inventory handling cost, production lead-time, delivery unit cost*, and *demand* were among the main criteria in that research. Lee (2005) in a research evaluated candidate places using the strength, weakness, and overall performance indexes applied to select the optimal location of distribution centers in logistics system design. The candidate places in that research were

Table 5.1 A comprehensive framework of sustainability criteria for evaluating liquid biofuels distribution centers

Category	Criteria	Sub-criteria	
Economic	Transportation and infrastructure	Logistics service provider	
		Connectivity to multi-modal transport	
			Port
			Highway and main road
			Airport
			Railway
		Quality and reliability of transportation modes	
		Distance to customer market	
		Distance to production facilities	
		Distance to supplier	
		Transportation cost	
		Extension transportation convenience	
		Density of shipping lines	
	Investment cost	Construction cost	
		Incentives	
		Land cost	
	Operational cost		
		Energy cost (e.g., water, electricity)	
		Labor cost	
		Tax structure	
		Climate condition cost	
		Handling cost Inventory cost	
	Economic risks	Earthquake possibility	
		Security	
	Export and import volume		
	Market related factors		
		Market size	
		Grow potential	
		Lead times and responsiveness	
	Warehouse facilities		
	Expansion possibility		

(continued)

Table 5.1 (continued)

Category	Criteria	Sub-criteria	
Social	Skilled labor		
	Policies of government		
	Cost of living in the region		
	Demand factors		
		Demand level	
		Demand volatility	
		Demand dispersion	
	Returnability		
	Potential impact factors related to the local social conditions		
	Human resource		
Environmental	Environmental impact		
	Hydrology condition		
	Impact on air pollution		

assessed by the criteria including *investment cost, expansion possibility, closeness to demand market, availability of acquirement material*, and *human resources.*

Løken and Botterud (2005) solved the location problem of distribution centers in an energy industry. To this end, first, the performance of alternatives in the five criteria was separately calculated through the multi-criteria utility theory and AHP methods, and then by comparison of the results of the two methods, the best location was selected. *Investment and operation costs, heat dump from plants to the environment, and NOx* and *CO_2 emissions* were the criteria used in that study. Wang and Kao (2007) in a research determined the best location of distribution centers in logistics companies. The rank of alternatives in that research was calculated by the fuzzy technique for order of preference by similarity to ideal solution (TOPSIS). *Climate conditions, demand quantity, expansion possibility, investment cost, labor force quality*, and *transportation availability* were among the criteria used to assess places in that research.

Lee and Lin (2008) identified the best location of international distribution centers in Asia using *strengths, weaknesses, opportunities,* and threats (SWOT) analysis and AHP in fuzzy environment. Fifteen criteria including *information abilities, transshipment time, port and warehouse facilities, port operation legal guarantee, export and import volume, density of shipping line, transshipment volume*, and *political, economic, society stability* were used to determine the performance of candidate places in that study. Anagnostopoulos et al. (2008) proposed a multi-criteria algorithm based on linguistic variables to determine the optimal location of distribution centers. The proposed algorithm works based on TOPSIS. In that research, the rank of alternatives was analyzed by six criteria including *expansion possibility, square measure of area, investment cost, availability of acquirement material, closeness to demand market*, and *human resources.*

In a study conducted by Ou and Chou (2009), the location of an international distribution center was determined based on six criteria including *market potential, infrastructure, transportation and distribution systems, cultural issues*, and *quality of services*. In that research, the weighted fuzzy factor rating system (FRS) was used as a methodology to evaluate places. Hu et al. (2009) suggested a hybrid methodology including the fuzzy set theory, simple additive weighing (SAW) method, and TOPSIS to select the best location of distribution centers. They solved this problem in a group decision-making environment. To this end, in that research, first, the weight of criteria including *price, market, transportation*, and *after service* was determined by SAW method, and then the rank of places was calculated by TOPSIS. Ji and Huailin (2009) employed genetic algorithm (GA) and AHP to determine the best location of logistics distribution centers. Based on the proposed method, at first, places with minimum cost were screened by GA, and then the rank of locations considering *environmental* and *service factors* was computed by AHP. Wang et al. (2010) in a research determined the best location of logistics distribution centers by using AHP. In that research, 16 criteria in 5 categories including *raw material availability, human resource, distribution network, availability of infrastructure*, and *close proximity to market* were used to assess alternatives' performance. The results of that paper indicated that *degree of proximity* in the *raw material availability* category was the main factor in selecting the best location of distribution centers. Demirel et al. (2010) in a study selected the optimal location of distribution warehouses for one of the major logistics companies in Turkey. They applied the Choquet integral method as a methodology in that research. To assess candidate places in that research, 16 criteria under 5 dimensions including *cost, labor characteristics, infrastructure, market*, and *environmental* are considered.

Kuo (2011) applied multi-criteria decision-making methods to select the best location of international distribution centers in Taiwan. In their study, first, the relationship between criteria was identified by decision-making trial and evaluation laboratory (DEMATEL), and then the rank of places was determined through three methods including TOPSIS, AHP, and analytic network process (ANP). *Imports and exports volume, port rate, location resistance, information abilities, port and warehouse facilities, port operation system, extension transportation convenience*, and *density of shipping line* were the criteria employed to assess alternatives. Awasthi et al. (2011) suggested fuzzy TOPSIS as a methodology to select the optimal location of urban distribution centers in a logistic company. Alternatives in that research were analyzed by 11 criteria such as *accessibility, connectivity to multimodal transport, cost, environmental impact, proximity to customers, resource availability*, and *possibility of expansion*. Using the axiomatic fuzzy set (AFS) and TOPSIS, Wang et al. (2012) ranked candidate places for urban distribution centers in a logistic company in China. Sixteen criteria categorized into six dimensions including *natural environment, transportation, business environment, candidate land, supply condition*, and *environmental impact* were used to analyze alternatives in that research.

Ashrafzadeh et al. (2012) determined the optimal location of distribution warehouses for an Iranian company using fuzzy TOPSIS. Alternatives in that research were assessed by 15 criteria such as *labor cost, transportation cost, handling cost, land cost, skilled labor, availability of labor force, proximity to customers and lead times*, and *responsiveness*. Chakraborty et al. (2013) proposed a hybrid methodology to determine the optimal location of distribution centers in a supply chain. The process of selection was carried out in two steps in that research. In this regard, first, the weight of criteria such as *access to infrastructure, reliability in operations, closeness to market, expert personnel availability*, and *earthquake possibility* was calculated by AHP.

Bouhana (2013) employed fuzzy theory and fuzzy preference relation as methodologies to select the optimal location of urban distribution centers. In that research, places were analyzed by ten criteria such as *accessibility, security, connectivity to multimodal transport, cost, environmental impact, proximity to customers, possibility of expansion*, and *quality of service*. Bagum and Rashed (2014) determined the optimal location of drug distribution centers in Bangladesh by using AHP. For this purpose, a framework of 17 criteria categorized into 5 dimensions including *economic, warehouse cost, market-related factors, transportation*, and *facility related factors* was suggested in that research. Based on the results of this research, *availability of transportation, suitability of land, market proximity, market size*, and *land cost* were weighted as the main criteria. Khan et al. (2015) determined the best location of food distribution centers in Pakistan. They used a hybrid methodology including rough set theory approach TOPSIS for determining the feasibility of candidate locations and ranking places, respectively. In that research 42 factors were considered such as *access to main roads, acceptance from district authorities, cost*, and *hazard risks*.

Neumüller et al. (2015) suggested ANP and process analysis method as methodologies to select the optimal location of distribution centers for a manufacturer company of confectionary in Germany. Places in that research were evaluated by criteria categorized into *risk, cost, opportunities*, and *benefits* dimensions. According to the results of that research, costs and risk were the main dimensions to select alternatives. Dey et al. (2016) in a study determined the optimal location of a distribution warehouse of a company located in India. To this end, they proposed a framework of 16 criteria categorized into *cost, infrastructure, market, work force*, and *environment* dimensions. Agrebi et al. (2017) proposed elimination and choice expressing reality (ELECTRE) as a methodology to solve the selection problem of logistics distribution centers. The candidate places in that research were evaluated by six criteria including *security, connectivity to multimodal transport, costs, proximity to customers, proximity to supplier*, and *conformance to sustainable freight regulations*. Onstein et al. (2020) using BWM provided insight into the weight of criteria involved in choosing the optimal spatial distribution structure. For this aim, they calculated the weights of 33 sub-criteria categorized into 7 dimensions. The results of that research indicate that logistics cost is the main factor.

As a result of the literature review, a comprehensive framework of sustainability criteria is presented in Table 5.1. By using the proposed framework, the utility of candidate places can be calculated in different location problems.

As presented in Table 5.1, in total there are 33 criteria which are categorized into economic, social, and environmental dimensions. From among the identified criteria, *transportation availability, land cost, market potential*, and *lead time* categorized

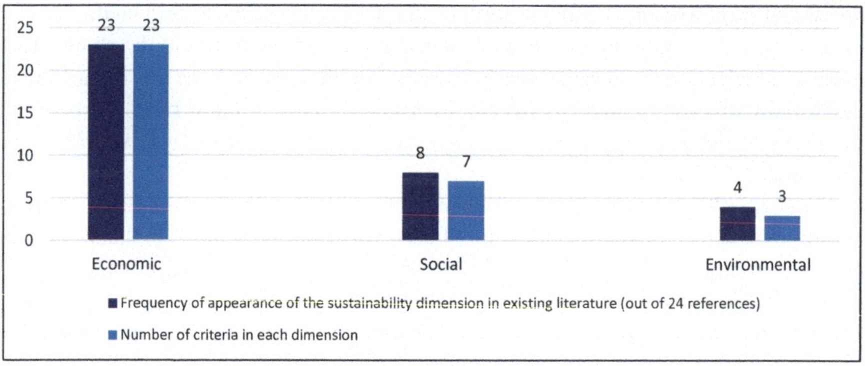

Fig. 5.2 The statistics of the papers reviewed in the three dimensions of sustainability

into the *economic* dimension and *skilled labor* from the *social* dimension are the ones with the highest frequency of appearance in existing literature. Figure 5.2 shows the number of sub-criteria and the papers cited the sub-criteria from the three dimensions.

The literature review also reveals that AHP is the most widely used method in the location problem of distribution centers. In this study we use BWM. Some advantages of BWM which justify its application as compared to AHP are mentioned as follows:

- The pairwise comparison provided by BWM is more consistent than that provided by AHP because BWM uses a systematic structure (using two opposite reference points) to collect data.
- The number of comparisons in BWM is less than that in AHP ($2n - 3$ against $n(n - 1)/2$)(Rezaei 2015).
- The process of weighting provided by BWM is easy to understand by the decision-maker/expert, which could lead to higher acceptability of the results by the decision-maker/expert (Rezaei 2015).

3 Proposed Methodology

In this section, a hybrid methodology, a BWM-set covering model, is presented to solve the location problem of bioethanol distribution centers. The proposed methodology deals with two objectives: (1) maximizing the utility of selected centers and (2) minimizing the cost allocated to establish distribution centers. In the proposed method, first of all, the utility of alternatives (the first objective function coefficients of variables) must be determined. For this purpose, calculating the performance of each alternative in the criteria presented in Table 5.1 is suggested. After determining the first objective function coefficients of variables, the best places for establishing bioethanol distribution centers are specified using set covering model. To solve the

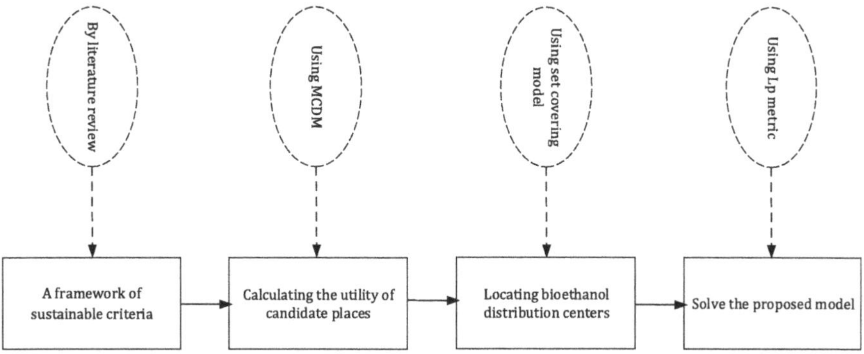

Fig. 5.3 The process of locating bioethanol distribution centers

suggested model, Lp-metric is used. This research methodology is summarized in Fig. 5.3.

3.1 Best-Worst Method

Best-worst method (BWM) as an MCDM method calculates the weight of criteria based on pairwise comparison. BWM has been used as a methodology to solve many major decision-making problems such as location selection (Kheybari et al. 2019a), technology evaluation (Kheybari et al. 2019a, b), energy management (van de Kaa et al. 2019), and sustainable supply chain management (Ahmad et al. 2017a, b; Wan Ahmad et al. 2016; Rezaei et al. 2019). To employ BWM there are five steps which are described as follows (Rezaei 2016):

1. Determine a set of decision criteria $\{c_1, c_2, \ldots, c_n\}$.
2. Identify the best (B) and the worst (W) criteria.
3. Determine the preference of the best over all the other criteria by a number from 1 to 9 (where 1 is "equally important" and 9 is "extremely more important"). Vector $A_B = (a_{B1}, a_{B2}, \ldots, a_{Bj}, \ldots, a_{Bn})$ is the result of steps 3 where a_{Bj} specifies the preference of indicator B over indicator j .
4. Determine the preference of all the criteria over the worst using the same scale we use in Step 3. $A_w = (a_{1W}, a_{2W}, \ldots, a_{jW}, \ldots, a_{nW})$ is the result of step 4 where a_{jW} shows the preference of indicator j over indicator W .
5. Compute the optimal weights $(w_1^*, w_2^*, \ldots, w_n^*)$.

   The optimal weights are calculated by minimizing the maximum absolute difference of $\{|w_B - a_{Bj}w_j|, |w_j - a_{jW}w_W|\}$ for all j which is translated into the following optimization problem:

$$\min \max_{j} \left\{ \left| w_B - a_{Bj} w_j \right|, \left| w_j - a_{jW} w_W \right| \right\}$$

such that:

$$\sum_{j=1}^{n} w_j = 1 \qquad w_j \geq 0, \quad \text{for all } j \tag{5.1}$$

Model (5.1) is converted into:
min ξ such that

$$\left| w_B - a_{Bj} w_j \right| \leq \xi, \quad \text{for all } j$$
$$\left| w_j - a_{jW} w_W \right| \leq \xi, \quad \text{for all } j$$
$$\sum_{j=1}^{n} w_j = 1 \qquad w_j \geq 0, \text{for all } j \tag{5.2}$$

Both $w^* = (w_1^*, w_2^*, \ldots, w_n^*)$ and ξ^* are the results of Model 2. w^* indicates the optimal weight of criteria and ξ^* shows the consistency of pairwise comparisons provided by respondents. If ξ^* is close to zero, it means that there is a high level of consistency in the pairwise comparison.

When there are sub-criteria in the proposed framework, the result of Model 2 for each level is called local weights. By multiplying the local weight of sub-criteria by the weight of their parent criteria the global weights are calculated. After calculating the global weight, the utility of alternative i, V_i, is computed by using Eq. 5.3.

$$V_i = \sum_{j} w_j^* u_{ij} \quad \text{for all } i \tag{5.3}$$

where u_{ij} is the normalized value of each alternative in terms of decision-making criteria calculated by Eqs. 5.4 or 5.5 for positive and negative criteria, respectively.

$$u_{ij} = \frac{x_{ij}}{\max_i \{x_{ij}\}} \quad \text{for all } i \text{ and } j \tag{5.4}$$

$$u_{ij} = \frac{\min_i \{x_{ij}\}}{x_{ij}} \quad \text{for all } i \text{ and } j \tag{5.5}$$

3.2 Set Covering Model

A set covering model is used in cases such as service facility locating problems, airline crews on a flight, assigning customers to delivery routes, and military logistics problems (Karasakal and Karasakal 2004; Lanza-Gutierrez et al. 2017). It is a zero-one linear programming model. Covering problems hold a central place in the location theory (Farahani and Asgari 2007). In these problems, we are given a set of demand points and potential sites for locating facilities. A demand point is claimed to be covered by a facility if it lies within a pre-specified distance of that facility. Covering problems are divided into two main classes, namely, set (total) covering problems, which cover all demand points with a minimum number of facilities, and maximal (partial) covering problems, which cover a maximum number of demand points with a fixed number of facilities (Church and Velle 1974). Based on the information provided in this section, the set covering model for the problem is as follows:

$$Max\ Z_1 = \sum_{i=1}^{m} V_i y_i \tag{5.6}$$

$$Min\ Z_2 = \sum_{i=1}^{m} y_i \tag{5.7}$$

Subject to:

$$\sum_{i=1}^{m} a_{di} y_i \geq 1 \qquad \text{for all } d \tag{5.8}$$

$$y_i \in \{0, 1\} \tag{5.9}$$

The first objective function (Eq. 5.6) represents the utility of the chosen places for establishing bioethanol distribution centers. As the number of centers decreases, the second objective function (Eq. 5.7) in the suggested model decreases the cost allocated to establish distribution centers. In this model, m is the number of candidate places for locating bioethanol distribution centers, and n is the number of places that should be covered by distribution centers. y_i is a binary variable; it is one if the candidate i is suitable for establishing bioethanol distribution centers and zero otherwise. V_i is the output of MCDM for candidate alternatives. The coverage matrix (a_{di}) is a binary parameter so that if the distance between the candidate place i and service applicant d is less than or equal to coverage radius, a_{di} is one and otherwise it is equal to zero. The coverage matrix is formed based on the coverage radius, and the coverage radius is the maximum distance to a center which can provide services for service applicants. In other words, it acts as its supporter. The constraint 8 presented in the model shows places satisfying coverage radius.

3.3 Lp-Metric

There are various assessment methods for multi-objective problems. Choosing a suitable method depends on the time limit, information, and preferences of a decision-maker. One of the easiest methods is the Lp-metric. In this method, the purpose is to minimize the deviation of ideal solutions from the existing objective functions (see Eq. 5.10):

$$\text{Min } L-P=\left\{\sum_{k=1}^{H}\gamma_k\left[\frac{f_k(y_k^*)-f_k(y_k)}{f_k(y_k^*)}\right]^p\right\}^{1/p} \tag{5.10}$$

where y_k^* shows the ideal solution in optimizing the kth objective and γ_k shows the importance (weight) of the kth objective ($\gamma_k > 0$). Moreover, $1 \leq p \leq \infty$ shows the amount of emphasis on the existing deviations so that if p goes up, the emphasis is greater on the deviations (Cao and Zhang 2009). After rewriting the coverage model, it is as the following:

$$\text{Min } L-P=\left\{\gamma_1\left[\frac{Z_1^*-Z_1}{Z_1^*}\right]^p+(1-\gamma_1)\left[\frac{Z_2^*-Z_2}{Z_2^*}\right]^p\right\}^{1/p} \tag{5.11}$$

Subject to:

$$\sum_{i=1}^{m} a_{di}y_i \geq 1 \qquad \text{for all } d \tag{5.12}$$

$$y_i \in \{0,1\} \tag{5.13}$$

An advantage of using this model is the normalization it does for the objective functions. That is, although the two objectives functions used in this study are of two different scales, the normalization done in the Lp model enables us to form a single-objective model.

4 Case Study

In this section, we employ the proposed methodology to determine the appropriate location of bioethanol distribution centers in Iran. To this end we first screen the criteria presented in Table 5.1. By screening the criteria, the discrimination power of decision criteria increases (Kheybari et al. 2019a). For this purpose, the opinion of six experts collected by a five-point Likert scale questionnaire is employed. After aggregating the questionnaires, due to existing approximate balance among the sub-criteria in each dimension of sustainability, the number of 3 (out of 5) is selected as a screen number (see Fig. 5.4). As can be seen from Fig. 5.4, for the dimension Environmental, no sub-criteria passes the cutoff point (3), which implies we keep this main criterion but no comparison is done for its sub-criteria.

In the next step of this research, BWM is employed to determine the weight of the selected criteria. In this regard, we collected the opinion of 15 experts through an online questionnaire designed based on BWM. All experts were identified through their profile on LinkedIn. The detail of all experts engaged in the weighting process is presented in Table 5.2. The result of BWM for the three dimensions of sustainability (i.e., economic, social, and environmental) is explained as follows.

Result of MCDM

In this part of chapter, we discuss the weight of relevant criteria and sub-criteria, the arithmetic mean of weights obtained from different experts by BWM. To clarify the dispersion of experts' opinion, we also calculate the standard deviation of the weight of criteria presented in Fig. 5.4.

According to experts' opinion, *economic, social, and environmental* aspects of candidate places have the maximum impact on bioethanol distribution centers in Iran, respectively (see Table 5.3). Unstable economic condition of Iran which increases the risk of investment and also lack of adequate infrastructure are two main reasons that justify the high weight of *economic* dimensions.

In the *economic* category, *investment cost* with the weight of 0.255 is identified as the most influential sub-criterion (see Table 5.4). Besides the high risk of investing, the high inflation rate in Iran and the existence of sanctions which limits the import of equipment to Iran are other two reasons that support the high weight of *investment cost*. In this category, *transportation and infrastructure*, *operational costs*, *market-related factors*, and *economic risks* are also other important criteria (see Table 5.4). The result of weighting analysis for the criteria categorized into social category indicates that the weight of *skilled labor* in the location selection of bioethanol distribution centers is more than the other two criteria (see Table 5.4). High unemployment rate of educated people in Iran could explain this high weight. *Human resources* and *cost of living* are also weighted as the second and third criteria in this category (see Table 5.4).

The high weight of *incentives* in comparison with *land cost* presented in Table 5.5 point out the undeniable role of government to facilitate the implementation of such centers in Iran. From among the six sub-criteria categorized into *transportation and infrastructure* category, *quality and reliability of transportation modes* with the weight of 0.238 is more important than the other five sub-criteria. Since infrastructure, which affects the quality of transportation service, is not the same in different places of Iran, candidate places that could support different types of transportation are considered with high priority. *Connectivity to multimodal transport, transportation costs, extension transportation convenience, logistics service provider*, and *density of shipping lines* are other important sub-criteria in this category (see Table 5.5). The high weight of *Lead time and responsiveness* categorized as sub-criterion into *market-related factors* category could be due to the large size of the country. In other words, there is a link between the location of distribution

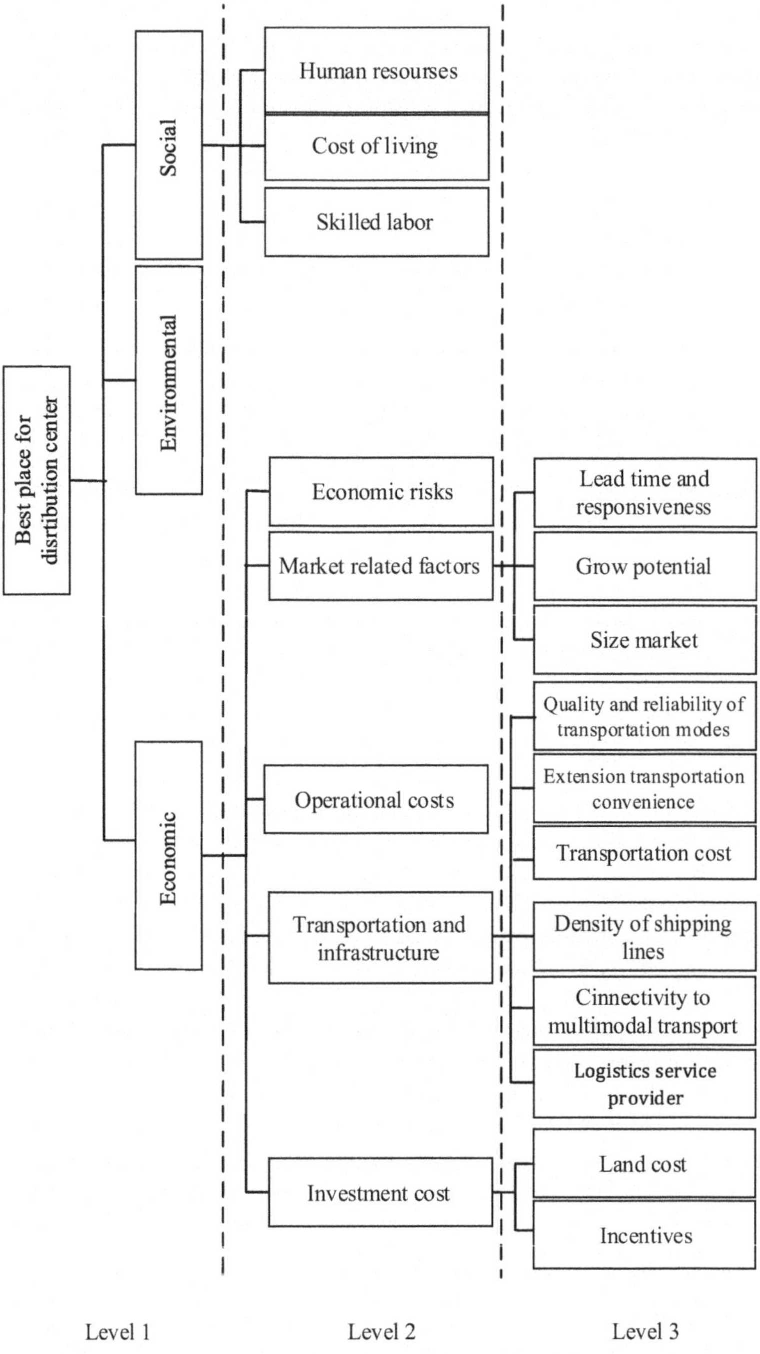

Fig. 5.4 Hierarchical tree of screened criteria

Table 5.2 The specification of experts used in this research

Respondents	For screening criteria	Average years of work experience	For weighting criteria	Average years of work experience
National Iranian Oil Company	–	–	7	14
Research Institute of Petroleum Industry	–	–	4	3.5
Renewable Energy and Energy Efficiency Organization	6	9	4	7.5

Table 5.3 Weights of the three dimensions of sustainability

Criteria	Weight	Standard deviation	Rank
Economic	0.382	0.143	1
Social	0.335	0.130	2
Environmental	0.283	0.129	3

Table 5.4 Weight of sub-criteria in level 2

Category	Sub-criteria	Weight	Standard deviation	Rank
Economic	Investment cost	0.255	0.115	1
	Transportation and infrastructure	0.238	0.121	2
	Economic risks	0.118	0.043	5
	Market-related factors	0.164	0.095	4
	Operational cost	0.225	0.078	3
Social	Skilled labor	0.378	0.148	1
	Cost of living	0.278	0.138	3
	Human resource	0.343	0.144	2

centers and *lead times* provided by each service provider. So, when there is a large area that should be covered by specific number of distribution centers, the candidate places that provide service to a greater number of service applicants are considered as high potential alternatives. Based on experts' opinion, *grow potential* and *market size* are the other two important sub-criteria in this category (see Table 5.5).

To calculate the utility score of each province, we calculate the global weight of sub-criteria presented in the last level of Fig. 5.4 (see Table 5.6).

As Table 5.6 indicates, the *environmental factor*, *skilled labor*, *human resource*, and *cost of living* are the first four sub-criteria that together account for about 60% of the total weight (see Table 5.6). After calculating the global weight of sub-criteria, the utility of alternatives, provinces of Iran, is determined based on Eq. 5.3. To this end, we collected the value of provinces of Iran (i.e., x_{ij} in Eqs. 5.4 and 5.5) with respect to the selected criteria from different databases including Statistical Center of Iran, the Ministry of Culture and Islamic Guidance, and Ministry of Petroleum. The result of this step is presented in Table 5.7.

Table 5.5 Weights of criteria in level 3

Criteria	Sub-criteria	Weight	Standard deviation	Rank
Investment cost	Incentives	0.636	0.287	1
	Land cost	0.364	0.287	2
Transportation and infrastructure	Logistics service provider	0.140	0.064	5
	Connectivity to multimodal transport	0.185	0.077	2
	Quality and reliability of transportation modes	0.238	0.120	1
	Transportation cost	0.161	0.073	3
	Extension transportation convenience	0.145	0.099	4
	Density of shipping lines	0.131	0.072	6
Market related factors	Market size	0.299	0.149	3
	Grow potential	0.349	0.136	2
	Lead time (LT) and responsiveness	0.352	0.141	1

Table 5.6 Global weights of sub-criteria

Criteria	Weight	Rank
Environmental factor	0.283	1
Skilled labor	0.127	2
Human resource	0.115	3
Cost of living	0.093	4
Operational cost	0.086	5
Incentives	0.062	6
Economic risks	0.045	7
Land cost	0.035	8
Lead Times and responsiveness	0.022	9
Grow potential	0.022	10
Quality and reliability of transportation modes	0.022	11
Market size	0.019	12
Connectivity to multimodal transport	0.017	13
Transportation cost	0.015	14
Extension transportation convenience	0.013	15
Logistics service provider	0.013	16
Density of shipping lines	0.012	17

Result of Set Covering Model Besides the utility of candidate places, the distance between the location of distribution centers and service applicants also affects the quality of service provided by distribution centers. Therefore, the geographical distribution should be also considered in the location selection of such centers. To satisfy this point, a multi-objective set covering model is applied in this research. We solve some examples to analyze the performance of the proposed model. To this end,

Table 5.7 Utility of provinces of Iran for establishing bioethanol distribution centers

Provinces of Iran	V_i	Rank
Ilam	0.0662	1
Tehran	0.0555	2
Kohgiluyeh and Boyer-Ahmad	0.0477	3
Khuzestan	0.0473	4
South Khorasan	0.0400	5
Razavi Khorasan	0.0359	6
Semnan	0.0347	7
Chaharmahal and Bakhtiari	0.0343	8
North Khorasan	0.0340	9
Kermanshah	0.0331	10
Fars	0.0314	11
Bushehr	0.0313	12
Isfahan	0.0309	13
Golestan	0.0304	14
Sistan and Baluchestan	0.0300	15
Ardabil	0.0295	16
Lorestan	0.0292	17
Yazd	0.0283	18
Kordestan	0.0283	19
East Azerbaijan	0.0273	20
Zanjan	0.0268	21
Kerman	0.0268	22
Hormozgan	0.0260	23
Ghom	0.0256	24
Guilan	0.0251	25
Hamedan	0.0250	26
West Azerbaijan	0.0249	27
Mazandaran	0.0249	28
Ghazvin	0.0247	29
Markazi	0.0224	30
Alborz	0.0221	31

we first evaluate the impact of coverage radius on the selected places, and then through some sensitivity analysis done on the weight of objective functions, the output of model is assessed. Note that (i) the value of a_{di} in all examples is calculated based on the distance between provinces of Iran, and (ii) we use p=1 in Eq. 5.11 for all examples.

As presented in Table 5.8, by increasing the coverage radius, both cost allocated to establish distribution centers (i.e., Z_2) and the utility of selected places are satisfied simultaneously. It means that when there is not limitation in the coverage radius of candidate places, the result provided by MCDM is to a great extent acceptable.

To evaluate the role of objective functions in the places selected for bioethanol distribution centers, we solve the model for coverage radius 500. As Table 5.9

Table 5.8 Result of model for different amount of coverage radius

Coverage radius	Selected places	Objective function
500	East Azerbaijan	0.79
	South Khorasan	
	Kerman	
	Kermanshah	
	Kohgiluyeh and Boyer-Ahmad	
	Golestan	
700	Razavi Khorasan	0.86
	Kordestan	
	Kerman	
	Kohgiluyeh and Boyer-Ahmad	
1000	Tehran	0.91
	Yazd	
1500	Kohgiluyeh and Boyer-Ahmad	0.95

Table 5.9 Sensitivity analysis done on the weight of objective functions

γ_1	0.1	0.2	0.3	0.4	0.5	0.6	0.7	0.8	0.9
Utility function (Z_1)	0.21	0.21	0.21	0.21	0.21	0.21	0.30	0.35	0.10
Cost function (Z_2)	6	6	6	6	6	6	7	8	31

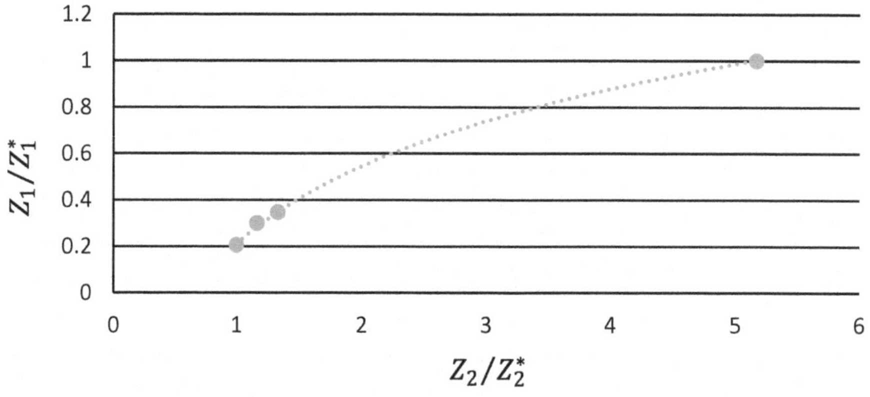

Fig. 5.5 Feasible area of objective functions for coverage radius 500

indicates, by increasing the weight of cost function (i. e. , $(1 - \gamma_1)$), the number of selected centers decreases. It means that the model tries to select the locations which could cover the demand of applicants, though they may not have high rank in the utility calculated by MCDM. On the other hand, when the weight of utility function increases (i. e. , γ_1), all service applicants receive high-quality service from high utility service provider.

The objective function space for different weight applied in Table 5.9 is drawn in Fig. 5.5. In brief, by determining an appropriate weight for objective functions, not only the utility of selected location is considered, but also the geographical distribution of service providers is also fulfilled with the minimum number of places.

5 Conclusion

The aim of this chapter was to propose a methodology to solve the location problem of bioethanol distribution centers. In this regard, a hybrid methodology, a BWM-set covering model, which determines most desirable places for distribution centers with minimum cost, was suggested. To calculate the utility of candidate places using BWM in the proposed method, a comprehensive framework of sustainability criteria was proposed by reviewing existing studies. The proposed framework provided useful information for experts about different aspects of bioethanol distribution centers.

Based on the information provided in the framework, economic and environment dimensions had the maximum and minimum criteria, respectively. The number of references citing the economic dimension was far more than those citing the other two aspects. By using the proposed framework and an MCDM method, such as BWM suggested in this study, the performance of candidate alternatives can be computed for any bioethanol distribution center. To select the optimal location of distribution centers, a set covering model was formulated. We also suggested Lp-metric as a method to find the solution of the proposed model.

The hybrid methodology was evaluated based on the data collected from Iran. The result of evaluation indicated that *environmental factor* has the maximum weight among all the criteria and Ilam has the highest utility to establish bioethanol distribution center in Iran

This chapter provides useful information for both decision-makers and scholars. Using the criteria presented in the framework, decision-makers can identify and resolve the shortcomings which adversely affect the supply of bioethanol. Furthermore, the methodology discussed in this study is useful to select the optimal location of distribution centers which should cover customers in a specific coverage radius, in different fields such as oil and chemical industries.

References

Agrebi, M., Abed, M., & Omri, M. N. J. J. o. A. T. (2017). ELECTRE I based relevance decision-makers feedback to the location selection of distribution centers. *Journal of Advanced Transportation, 2017*, Article ID 7131094.

Ahmad, W. N. K. W., Rezaei, J., Tavasszy, L. A., & de Brito, M. P. (2016). Commitment to and preparedness for sustainable supply chain management in the oil and gas industry. *Journal of Environmental Management, 180*, 202–213.

Ahmad, N. K. W., de Brito, M. P., Rezaei, J., & Tavasszy, L. A. (2017a). An integrative framework for sustainable supply chain management practices in the oil and gas industry. *Journal of Environmental Planning and Management, 60*(4), 577–601.
Ahmad, W. N. K. W., Rezaei, J., Sadaghiani, S., & Tavasszy, L. A. (2017b). Evaluation of the external forces affecting the sustainability of oil and gas supply chain using Best Worst Method. *Journal of Cleaner Production, 153*, 242–252.
Anagnostopoulos, K., Doukas, H., & Psarras, J. J. E. S. w. A. (2008). A linguistic multicriteria analysis system combining fuzzy sets theory, ideal and anti-ideal points for location site selection. *Expert Systems with Applications, 35*(4), 2041–2048.
Ashrafzadeh, M., Rafiei, F. M., Isfahani, N. M., & Zare, Z. J. I. J. O. C. R. I. B. (2012). Application of fuzzy TOPSIS method for the selection of warehouse location: A case study. *Journal of Contemporary Research in Business, 3*(9), 655–671.
Awasthi, A., Chauhan, S. S., Goyal, S. K. J. M., & Modelling, C. (2011). A multi-criteria decision making approach for location planning for urban distribution centers under uncertainty. *Mathematical and Computer Modelling, 53*(1-2), 98–109.
Bagum, N., & Rashed, C. A. A. J. R. O. G. M. (2014). Multi-criteria analysis for the selection of location for distribution center using analytical hierarchy process. *Review of General Management, 20*, 2.
Ben-Iwo, J., Manovic, V., & Longhurst, P. (2016). Biomass resources and biofuels potential for the production of transportation fuels in Nigeria. *Renewable and Sustainable Energy Reviews, 63*, 172–192.
Bhushan, N., & Rai, K. (2007). *Strategic decision making: Applying the analytic hierarchy process*. Cham: Springer Science & Business Media.
Bouhana, A., Chabchoub, H., Abed, M., & Fekih, A. (2013). A multi-criteria decision making approach based on fuzzy theory and fuzzy preference relations for urban distribution centers' location selection under uncertain environments. In *2013 international conference on advanced logistics and transport* (pp. 556–561). IEEE.
Cao, F., & Zhang, R. (2009). The errors of approximation for feedforward neural networks in the Lp metric. *Mathematical and Computer Modelling, 49*(7), 1563–1572.
Chakraborty, R., Ray, A., & Dan, P. J. I. J. o. I. E. C. (2013). Multi criteria decision making methods for location selection of distribution centers. *International Journal of Industrial Engineering Computations, 4*(4), 491–504.
Chan, F., & Chung, S. J. T. I. J. o. A. M. T. (2004). Multi-criteria genetic optimization for distribution network problems. *International Journal of Advanced Manufacturing Technology, 24*(7-8), 517–532.
Chen, C.-T. J. F. s., & systems. (2001). A fuzzy approach to select the location of the distribution center. *Fuzzy sets and systems, 118*(1), 65–73.
Church, R., & Velle, C. R. (1974). The maximal covering location problem. *Papers in regional science, 32*(1), 101–118.
Demirel, T., Demirel, N. Ç., & Kahraman, C. J. E. S. w. A. (2010). Multi-criteria warehouse location selection using Choquet integral. *Expert Systems with Applications, 37*(5), 3943–3952.
Dey, B., Bairagi, B., Sarkar, B., Sanyal, S. K. J. I. J. o. M. S., & Management, E. (2016). Warehouse location selection by fuzzy multi-criteria decision making methodologies based on subjective and objective criteria. *International Journal of Management Science and Engineering Management, 11*(4), 262–278.
Farahani, R. Z., & Asgari, N. (2007). Combination of MCDM and covering techniques in a hierarchical model for facility location: A case study. *European Journal of Operational Research, 176*(3), 1839–1858.
Hu, Y., Wu, S., & Cai, L. (2009, April). Fuzzy multi-criteria decision-making TOPSIS for distribution center location selection. In *2009 international conference on networks security, wireless communications and trusted computing* (Vol. 2, pp. 707–710). IEEE.

Ji, L., & Huailin, D. (2009, July). Research on logistics distribution center location problem based on genetic algorithm and AHP. In *2009 4th international conference on Computer Science & Education* (pp. 213–217). IEEE.

Karasakal, O., & Karasakal, E. K. (2004). A maximal covering location model in the presence of partial coverage. *Computers & Operations Research, 31*(9), 1515–1526.

Kemp, K. (2008). *Encyclopedia of geographic information science*. London: Sage.

Khan, C., Anwar, S., Bashir, S., Rauf, A., Amin, A. J. J. o. I., & Systems, F. (2015). Site selection for food distribution using rough set approach and TOPSIS method. *Journal of Intelligent & Fuzzy Systems, 29*(6), 2413–2419.

Kheybari, S., Kazemi, M., & Rezaei, J. (2019a). Bioethanol facility location selection using best-worst method. *Applied Energy, 242*, 612–623.

Kheybari, S., Rezaie, F. M., & Rezaei, J. (2019b). Measuring the importance of decision-making criteria in biofuel production technology selection. *IEEE Transactions on Engineering Management*. https://doi.org/10.1109/TEM.2019.2908037.

Kuo, M.-S. J. E. S. w. A. (2011). Optimal location selection for an international distribution center by using a new hybrid method. *Expert Systems with Applications, 38*(6), 7208–7221.

Lanza-Gutierrez, J. M., Crawford, B., Soto, R., Berrios, N., Gomez-Pulido, J. A., & Paredes, F. (2017). Analyzing the effects of binarization techniques when solving the set covering problem through swarm optimization. *Expert Systems with Applications, 70*, 67–82.

Lee, H.-S. (2005, August). A fuzzy multi-criteria decision making model for the selection of the distribution center. In *International conference on natural computation* (pp. 1290–1299). Berlin, Heidelberg: Springer.

Lee, K.-l., & Lin, S.-c. J. I. S. (2008). A fuzzy quantified SWOT procedure for environmental evaluation of an international distribution center. *Information Sciences, 178*(2), 531–549.

Løken, E., & Botterud, A. (2005). *Planning of mixed local energy distribution systems: A comparison of two multi-criteria decision methods*. Paper presented at the 28th Annual IAEE International Conference, Taipei, Taiwan.

Neumüller, C., Kellner, F., Gupta, J. N., & Lasch, R. J. I. J. o. P. R. (2015). Integrating three-dimensional sustainability in distribution centre selection: The process analysis method-based analytic network process. *International Journal of Production Research, 53*(2), 409–434.

Onstein, A. T., Ektesaby, M., Rezaei, J., Tavasszy, L. A., & van Damme, D. A. (2020). Importance of factors driving firms' decisions on spatial distribution structures. *International Journal of Logistics Research and Applications, 23*(1), 24–43.

Ou, C.-W., & Chou, S.-Y. J. E. s. w. a. (2009). International distribution center selection from a foreign market perspective using a weighted fuzzy factor rating system. *Expert systems with applications, 36*(2), 1773–1782.

Prasad, S., Dhanya, M., Gupta, N., & Kumar, A. (2012). Biofuels from biomass: A sustainable alternative to energy and environment. *Biochemical and Cellular Archives, 12*(2), 255–260.

Rezaei, J. (2015). Best-worst multi-criteria decision-making method. *Omega, 53*, 49–57. https://doi.org/10.1016/j.omega.2014.11.009.

Rezaei, J. J. O. (2016). Best-worst multi-criteria decision-making method: Some properties and a linear model. *Omega, 64*, 126–130.

Rezaei, J., Papakonstantinou, A., Tavasszy, L., Pesch, U., & Kana, A. (2019). Sustainable product-package design in a food supply chain: A multi-criteria life cycle approach. *Packaging Technology and Science, 32*(2), 85–101.

Saini, J. K., Agrawal, R., Satlewal, A., Saini, R., Gupta, R., Mathur, A., & Tuli, D. (2015). Second generation bioethanol production at high gravity of pilot-scale pretreated wheat straw employing newly isolated thermotolerant yeast Kluyveromyces marxianus DBTIOC-35. *RSC Advances, 5*(47), 37485–37494.

Santibañez-Aguilar, J. E., González-Campos, J. B., Ponce-Ortega, J. M., Serna-González, M., & El-Halwagi, M. M. (2014). Optimal planning and site selection for distributed multiproduct biorefineries involving economic, environmental and social objectives. *Journal of Cleaner Production, 65*, 270–294.

van de Kaa, G., Fens, T., & Rezaei, J. (2019). Residential grid storage technology battles: A multi-criteria analysis using BWM. *Technology Analysis & Strategic Management, 31*(1), 40–52.

Wang, Y.-J., & Kao, C.-S. (2007). *Applying fuzzy multiple-criteria decision-making method to select the distribution center.* Paper presented at the fskd.

Wang, M.-H., Lee, H.-S., Chu, C.-W. J. I. J. O. I. C., & Information, & Control. (2010). Evaluation of logistic distribution center selection using the fuzzy MCDM approach. *International Journal of Innovative Computing, Information and Control, 6*(12), 5785–5796.

Wang, Y., Ma, X.-l., Wang, Y.-h., Mao, H.-j., & Zhang, Y. (2012). Location optimization of multiple distribution centers under fuzzy environment. *Journal of Zhejiang University Science A, 13*(10), 782–798.

Yoruklu, H. C., Koroglu, E. O., Demir, A., & Ozkaya, B. (2019). The electromotive-induced regulation of anaerobic fermentation: Electrofermentation. In *Microbial electrochemical technology* (pp. 739–756). Elsevier.

Chapter 6
Multiple Criteria Decision Analysis to Assess Urban and Territorial Transformations: Insights from Practical Applications

I. M. Lami, M. Bottero, and F. Abastante

Abstract This chapter aims at providing concrete reflections about the use of the Multiple Criteria Decision Analysis (MCDAs) in urban and territorial decision processes. The reflections provided by this chapter come from a critical analysis of different case studies faced by the Turin research group (Politecnico di Torino, InterUniversity Department DIST) composed by the authors, in some case in collaboration with other academics. Over the last decade in fact, the authors applied a number of MCDA to case studies of different nature in order to provide answers to decision-making problems in urban and territorial planning realms, towards a sustainable development. Specifically, the case studies analysed in this chapter refer to seven contexts: (i) infrastructural transport planning strategies; (ii) location of undesirable facilities; (iii) strategic urban planning; (iv) urban energy retrofitting; (v) real estate investments; (vi) cultural adaptive reuse of abandoned buildings; and (vii) environmental systems. The aforementioned case studies cover different geographical scales of intervention (local, national and transnational) offering here the opportunity to reflect around the use of MCDA as Analytic Network Process (ANP), ANP and Spatial Decision Support Systems (SDSS), Dominance Based Rough-Sets Approach (DRSA), Measuring Attractiveness by a Categorical Based Evaluation Technique (MACBETH), Preference Ranking Organization METHod for Enrichment of Evaluations (PROMETHEE), CATegorization by Similarity-Dissimilarity (CAT-SD) and Elimination Et Choix Traduisant La Realité (ELECTRE). The applications presented show that the MCDA can be a useful support to the decision-makers in order to structure the decision process in exam, characterized by a plurality of stakeholders with different interests, powers and goals. In particular, starting from the case studies, the authors highlight the applicability and the decision-making relevance of the different MCDA.

I. M. Lami (✉) · M. Bottero · F. Abastante
Politecnico di Torino, Inter-University Department of Regional and Urban Studies and Planning (DIST), Turin, Italy
e-mail: isabella.lami@polito.it

J. Rezaei (ed.), *Strategic Decision Making for Sustainable Management of Industrial Networks*, Greening of Industry Networks Studies 8,
https://doi.org/10.1007/978-3-030-55385-2_6

Keywords Multiple Criteria Decision Analysis · Urban and territorial transformations · Infrastructural transport planning · Urban planning · Urban energy retrofitting

1 Introduction

Territorial transformations refer to complex systems (Simon 1960) composed by a number of interacting aspects characterized by dynamics which cannot be faced separately. It has been generally agreed that MCDA can provide a useful support in order to structure decision processes related to territorial transformations, allowing several aspects to be considered in a complex situation (Roy and Bouyssou 1993; Figueira et al. 2005). The MCDAs help the decision-makers (DMs) considering qualitative and quantitative aspects and different point of views and integrate different options. However, as the MCDA are countless, it is necessary to deeply reflect on the most suitable method for the decision context in exam (Roy and Slowinski 2013).

The chapter illustrates a selection of methodologies applied to territorial and urban transformation projects, where the "sustainability question" takes on a socio-technical character, challenging the experts to provide the most appropriate methodology to support the DM. It is important to highlight how, in this specific context, the "most appropriate" sometimes means the one that allows a simpler understanding by the DM and the more communicable one, in order to allow the biggest participation in the decision process.

The reflections provided by this chapter come from a critical analysis of different case studies faced by the Turin research group (Politecnico di Torino, InterUniversity Department DIST) composed by the authors, in some case in collaboration with other academics. Specifically, the case studies analysed in this chapter refer to seven contexts: (i) infrastructural transport planning strategies; (ii) location of undesirable facilities; (iii) strategic urban planning; (iv) urban energy retrofitting; (v) real estate investments; (vi) cultural adaptive reuse of abandoned buildings; and (vii) environmental systems. The aforementioned case studies cover different geographical scales of intervention (local, national and transnational) offering here the opportunity to reflect around the use of several MCDA, such as Analytic Network Process (ANP), ANP and Spatial Decision Support Systems (SDSS), Dominance Based Rough-Sets Approach (DRSA), Measuring Attractiveness by a Categorical Based Evaluation Technique (MACBETH), Preference Ranking Organization METHod for Enrichment of Evaluations (PROMETHEE), CATegorization by Similarity-Dissimilarity (CAT-SD) and Elimination Et Choix Traduisant La Realité (ELECTRE).

The chapter is divided in four sections. After the introduction, the second section is aimed at giving a synthetic overview of the various MCDA used, with a specific focus on the decision-making and sustainability relevance. The third one briefly describes seven case studies, following a tripartite scheme (decision problem,

MCDA method, results and role played by the MCDA). The last section summarized the conclusions.

2 Urban and Territorial Transformation

The way in which urban regeneration is developed and implemented in Europe has changed over the decades since the post-war reconstruction and building boom. These modifications have also led to a change in the role of evaluation.

Between the 1960s and the late 1980s, territorial transformations were carried out following specific plans executed without closely verifying a fit between forecasts (made by planners) and the demand of the market (Healey 2007). Despite this consolidated dynamic, the planned transformations were successful since the cities were growing as well as the population and the Public Administrations (PA) were facing a financial well-being (Secchi 1984).

Due to an erosion of public incentives, the PA started to face a crisis during the 1990s that in turn caused a decrease of the real estate transactions and of the DM's power in controlling and guiding urban and territorial transformations. This crisis led to the need for public-private partnerships (Ruegg 1994) increasing the number of players involved in the decision-making processes including not simply those directly affected by the state of the area but also those who practically realize the transformation (e.g. owners, investors) and intermediate actors (e.g. real estate agents) (Faludi 1973).

The main consequence is that urban transformation operations are fragmented driving processes, with different time perspectives for operators. Each approach requires a specific metric to make the results achieved objectively measurable and comparable. The challenge is to coordinate different operations, which have different potential in terms of time, profitability and values (Lami 2019).

Moreover, nowadays many cities need to face new challenges. The social and economic changes, which are not aspatial also according to sustainability (Lami et al. 2014), concern material and immaterial aspects deeply affecting urban and territorial transformations. This concept of sustainable development, understood as the "development that meets the needs of the present without compromising the ability of future generations to meet their own needs" (Bruntland 1987), has been recently stressed out by the 2030 Agenda for Sustainable Development (sustainabledevelopment.un.org). The latter provides strategic directions to support the future development by identifying 17 Sustainable Development Goals (SDGs) considered as urgent call for actions in a global partnership.

In particular, the SDG11 "Make cities and human settlements inclusive, safe, resilient and sustainable" has become a pivotal attempting to analyse and solve the interdependences between environment, development and economic growth allowing the quality and efficiency of urban and territorial transformations to be considered.

The management of such complex territorial transformation processes requires a "governance of knowledge" (Mondini 2008) able to maximize the objectives of the biological system (resilience, bio-productivity), of the economic system (the fulfilment of basic needs, the strengthening of equity and the increase of goods and services) as well as the objectives related to the social system (cultural difference, institutional sustainability, justice and participation) (Barbier 1987).

Against this challenge, the MCDAs are suitable to reason with different stakeholders investigating a number of point of views and "needs" which are no longer clear, shared and declared. According to this perspective, the discussion around a territorial transformation always implies understanding if the territory needs for a specific route of the railway line, a specific waste treatment technology, a new building or the adaptive reuse of existing heritage, to satisfy our needs.

In this context, territorial transformation has to define the appropriate policy to tackle the interactions among the problems and the nexuses that led to a decline of the cities (De Magalhaes 2015).

3 MCDA and Case Studies

An urban and territorial transformation could be seen as a "search to balance needs, institutional and financial constraints and market responses, within a perspective of sustainability" (Lami 2014, p. 89).

In this perspective, the evaluation approaches progressively try to consider the complexity of the context, moving from an approach mainly based on the analysis of the urban/territorial factors and the real estate value to a more integrate approach, in which not only the spatial and the financial aspects of the project are considered but also the social implications and the environmental effects. Facing the new trends in the context of public participation at a European level, it is necessary to be more inclusive in the evaluation process, considering the use of specific tools enabling the involvement of the population in the decision process and to take the different opinions into account (Lami 2014).

Starting from this perspective, we have created a summary table of the cases hereinafter illustrated, with their main characteristics, in order to emphasize the decision-making relevance of the method and the sustainability relevance of the case study (Table 6.1).

In the first case study, concerning the evaluation of Corridor 24 with a mixed method combining ANP, Interactive Maps and Strategic Assessment (Abastante et al. 2014), the decision-making relevance of the method is extreme because, from the methodological point of view, the experiment represented a groundbreaking activity. In order to help the actors involved to understand spatial issues, each question of ANP was supported by the visualization of the symbolic positioning of the expected effects. The dynamic maps were displayed to participants showing, in real-time, the behaviour of maps if the given weight was 1 or 9 for one element or another. The workshops proved to be able to consolidate and enhance the

Table 6.1 Interpretation of the case studies

		CASE STUDIES							
		Infrastructural projects	Waste management	Strategy planning	Energy planning	Cultural heritage	Built environment	Environmental systems	
		Corridor 24 strategies	**Urban solid waste**	**Redevelopment of a district in Brussels**	**Net Zero Energy District (NZED)**	**Adaptive reuse in Turin**	**Real estate portfolio**	**Requalification of an abandoned quarry**	**Decision-making relevance**
MCDA methodologies	**ANP + Visual.**	X							Extreme
	DRSA		X						High
	MACBETH			X					Medium
	PROMETHEE				X				Medium
	CAT-SD					X			Extreme
	NAROR						X		High
	ELECTRE							X	Extreme
Spatial Scale		International	Local	Local	Local	Local	Local	Regional	
Sustainability relevance		High	High	Medium	High	High	Medium	Extreme	

collaboration between the participants, because they were guided in understanding where their choice might fall, so as to evaluate step by step the importance of their response. The combination of different inputs and modes of joint interaction created the right milieu for the creation of knowledge among DMs and researchers in the fields of transport, economics, environment and spatial planning.

In the second case study, which concerns the choice of the most suitable location for a Municipal Solid Waste Plant (MSWP) in the Province of Torino, Italy (Abastante et al. 2012), the decision-making relevance (of the method) is high, the DRSA offers a useful tool for reasoning about the data involved in the decision problem at hand, and it is suitable to elicitate the DM's preferences and to support them by explaining and justifying the final choice basing on easily understandable decision rules. As for the sustainability relevance (of the case study), it is high, because the problem of the location of MSWP is an intrinsically complex problem involving interconnected elements as social, economic and environmental. In particular, the citizens usually show phenomena as NIMBY (Not-In-My-Back-Yard), NOTE (Not-Over-There-Either), LULU (Locally-Unacceptable-Land-Use) and BANANA (Build-Absolutely-Nothing-Anywhere-Near-Anything). For this reason, the DMs need to able to justify their choices concerning the location of disposal sites through systematic, transparent and sustainable processes.

The third case study is a simulated academic process for the transformation of the Tür und Taxis district in Brussels (Belgium) (Abastante and Lami 2018), subject to a discussion from the Municipality of Brussels. The decision-making relevance (of the method) is medium; the stakeholders are supported in constructing numerical scales basing on qualitative comparisons in terms of differences of attractiveness overcoming the difficulty of representing personal preferences using numbers (Bana e Costa and Chargas 2004). Accordingly, the method has the ability to inspire discussion and thoughts rather than providing optimal answers to a problem. The advantage provided by the constant consistency checks of the relative ambiguity of possible answers can become a double-edge sword if applied in real-time focus groups. In fact, the risk is to extend too much the available time of the discussion, which is a huge problem mainly with political stakeholders. The sustainability relevance (of the case study) is medium. The physical and social fragmentation of cities is at the core of huge international sustainability discussion. Helping solving this issue will contribute to reduce the social exclusion, the growth of violence and the loss of social identity through more transparent and sensible decision processes.

The fourth case study concerns the application of the PROMETHEE method for the evaluation of alternative strategies for the creation of a Net Zero Energy District (NZED) in Turin (Becchio et al. 2017). As far as the decision-making relevance is concerned, the PROMETHEE method proved to be easy and reliable for the DMs who are able to understand the method itself and the concepts involved since they are represented by meaningful parameters, such as physical units and economic values. Furthermore, PROMETHEE method supports multi-actors and group decision-making, and it provides a useful tool for debate and consensus building. With reference to the sustainability relevance, the case study is interesting as the attention towards the urban scale in energy planning requires the consideration of the full range of aspects involved (i.e. environmental, economic and social dimensions).

The fifth case study is related adaptive reuse for cultural and art purposes of underused sites and buildings located in Turin, Italy (Costa et al. 2019). The decision-making relevance (of the method) is extreme since the DM is involved in the definition of the categories, in the selection of the actions and the reference actions, in the identification of the criteria and the criteria scale construction, in assigning criteria weights and in the identification of possible criteria interactions. It is a real co-constructive procedure for the whole process. The sustainability relevance (of the case study) is high, because the adaptive reuse of the abandoned building is a crucial aspect in order to avoid the environmental costs of the construction of new buildings and the urban sprawl.

In the sixth case study the decision-making relevance is high as the application is related to the development of the NAROR method for supporting the management of a real estate portfolio (Bottero et al. 2016). In this case, the DMs give preference information in a very simple way by comparing alternative or criteria from which to elicit parameters compatible with these preferences. As far as the sustainability relevance is concerned, this could be judged as medium because the evaluation mainly considers the socio-economic aspects of the problem.

The final case considers the application of an extension of the ELECTRE III with interaction to an environmental decision problem (Bottero et al. 2015). The decision-making relevance is extreme due to the novelty of the application. In the study, the method proved to be very appropriate and able to highlight and build robust conclusions, considering the existence of some arbitrariness in the definition of weights and interaction coefficients. Also, the sustainability relevance is high as the evaluation considers the existence of possible interactions between pairs of criteria. This is fundamental in the context of sustainability assessment where the substitution principle cannot be suitable and where it is necessary to assess the levels of synergy or redundancy among the main dimensions (environmental, social and economic).

4 Applications

4.1 Infrastructural Projects: Application of Analytic Network Process and Interactive Maps to the Genoa-Rotterdam Corridor

4.1.1 Decision Problem

The case study is related to the evaluation of Corridor 24 (Genoa-Rotterdam) alternative development strategies, and it has been analysed in the framework of an Interreg IVB NWE Project, called "Code24" (Abastante et al. 2014). The Code24 Project aimed at identifying a shared spatial and infrastructural sustainable strategy for the regions connected through this infrastructure of strategic European importance. Based on a problem-oriented and not on geographical approach, Corridor 24 has been divided into nine subregions, which present analogue problem backgrounds; for each region, alternative strategies have been explored and drafted. The key components of the strategies reflected the structure of the MCDA adopted as well (more precisely the clusters and nodes of the Analytic Network Process model).

4.1.2 MCDA Method

A Collaborative Assessment Workshop has been designed in order to allow different forms of interaction between the participating actors and enabling them to contributing referring to both their knowledge as local stakeholders and experts' skills. The assessment foresees, in fact, four different stages of interaction with the content materials and the other participants:

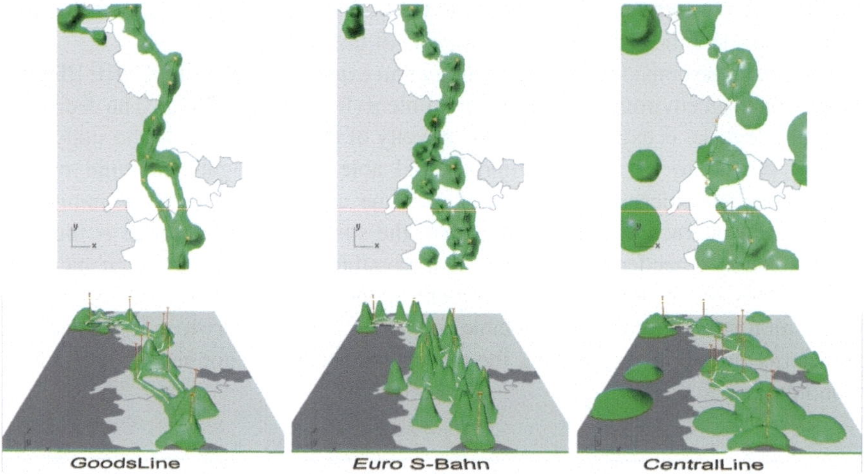

Fig. 6.1 Example of InViTo maps provided during the workshop. (Source: Abastante et al. 2014)

- Remote individual work.
- Structured sessions steered by a moderator and facilitators.
- Semi-structured sessions with the help of a facilitator.
- Unstructured and open discussion.

During the different sessions, the participants are asked not only to evaluate and provide feedback to materials presented by the organizers but also encouraged to present their own contributions and ideas.

The first step of the assessment has been structured through an approach integrating Analytic Network Process (ANP, Saaty 2005) and the Interactive Visualization Tool (InViTo), test visualization tools able to support the decision-making in real time, in order to create a shared basis for generating discussion (Lami et al. 2014). A single network ANP model was developed in order to generate a free and open discussion among the stakeholders involved. The decision problem in exam has been divided into five clusters (namely, economic development, spatial development, rail operation, environment and logistics). Each cluster has been divided in turn into elements (or nodes) representing the specific aspects of the decision problem (Fig. 6.1) and visualized in real time with InViTo (Pensa et al. 2013).

Mention has to be made to the fact that in this application, the ANP is not used as a method to determine a priority list of the different alternatives in the decision problem. The ANP model here is applied as a structured procedure that is able to support the assessment of the identification of the considered key development factors in order to come to a decision. In this sense, it is a rather rare application of ANP technique because the obtained results are not an order of alternatives but a sorting of criteria (Abastante and Lami 2013; Bottero and Lami 2010).

The second step was constituted by a Collaborative Assessment assuming the perspective of strategy design processes as peer learning practice. The Collaborative

Assessment Workshop regarding the whole Rotterdam-Genoa corridor has been conducted in two sessions of one and half days, but it came at the end of a larger preparatory process, composed of meetings, focus groups and workshops. The preparation has been important to generate the right collaboration climate for the implementation of the procedure and to allow the participants to be familiar with the issues and the material, which would then have been object of the assessment.

4.1.3 Results and Role Played by the MCDA

The aim of the Collaborative Assessment Procedure was to accompany the development of a shared position by the CODE24 project partnership regarding the most relevant issues affecting the future corridor's development. The assessment phase has therefore been an interactive process. It promoted an open and intensive discussion among the partners and other important stakeholders on the spatial and infrastructure development of the Rotterdam-Genoa corridor, in different regions and at interregional level.

From the methodological point of view the experiment represented a groundbreaking activity. It provided the right environment for testing innovative instruments and tools and collects important feedback on their performances. In particular, it seems that the combination of different inputs and modes of joint interaction is able to support a productive discussion also when dealing with extremely complex and challenging tasks. The workshops proved to be able to consolidate and enhance the collaboration between the participants, in spite of limited time and the difficulties arising from the different backgrounds of the stakeholders. The most significant sign of the action's success was the willingness of the involved partners to continue with the joint, collaborative work in the coming future.

4.2 Waste Management: The Role of the Dominance Base Rough Sets Approach for a Waste Disposal

4.2.1 Decision Problem

The decision problem presented concerns the choice of the most suitable location for a MSWP in the Province of Torino (Italy).

The research is based on an environmental analysis developed by the Provincial Administration (ATO-R 2007) highlighting 39 sites as potentially suitable for the location of the MSWP (Fig. 6.2). In the study, the 39 sites are analysed on the basis of qualitative and quantitative attributes according to specific indicators as presence of population, water table depth, vulnerability, valuable crops, number of farms, land use capacity, interference with traffic and operating costs (Abastante et al. 2012).

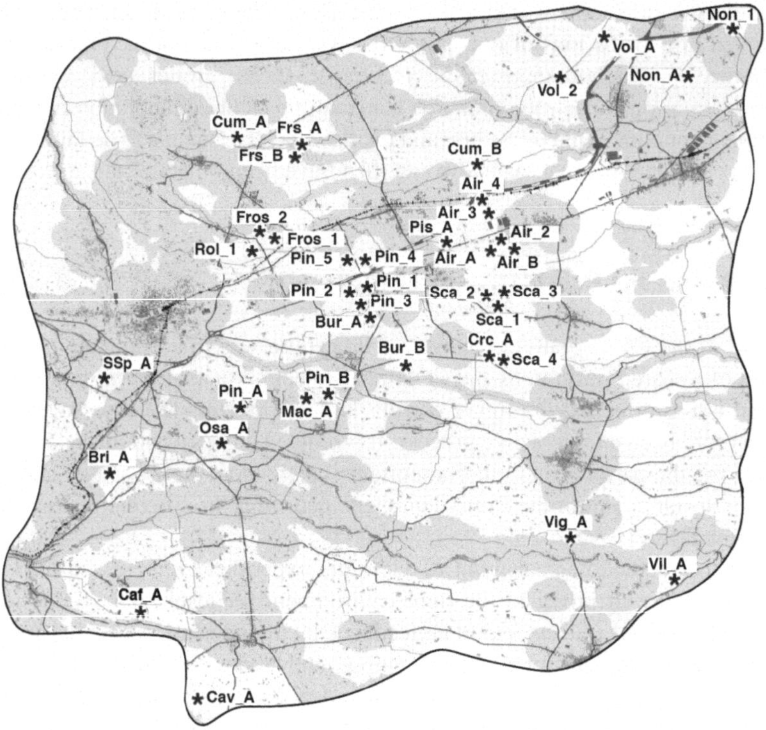

Fig. 6.2 Location of the considered sites. (Source: Abastante et al. 2012)

4.2.2 MCDA Method

In order to tackle the problem in exam, we applied an interactive Dominance-based Rough Set Approach (DRSA, Greco et al. 2001, 2008a). This method is based on the use of a preference model expressed in terms of easily understandable "*if* ... *then*" decision rules. The rules are induced from a preference information given by the DM in terms of indication of relatively good actions from a given sample called exemplary decision. The method consists of organizing the search of the most preferred action by alternating stages of calculation and dialogue. The calculation stages aim at inducing decision rules and applying them to new actions. During the dialogue stages, the DMs are called to react to the proposal revealing their preferences sorting the alternative as *good* or *bad*, which are in turn considered in the next stages of calculation. The procedure stops when the DMs find a satisfactory action (Fig. 6.3).

The case study in exam required four interactions with the DM. Apart from the first interaction, where only a subset of candidate sites has been considered, the number of *good* sites decreased from 39 to 23 (second interaction), to 6 (third interaction), to 2 (fourth interaction).

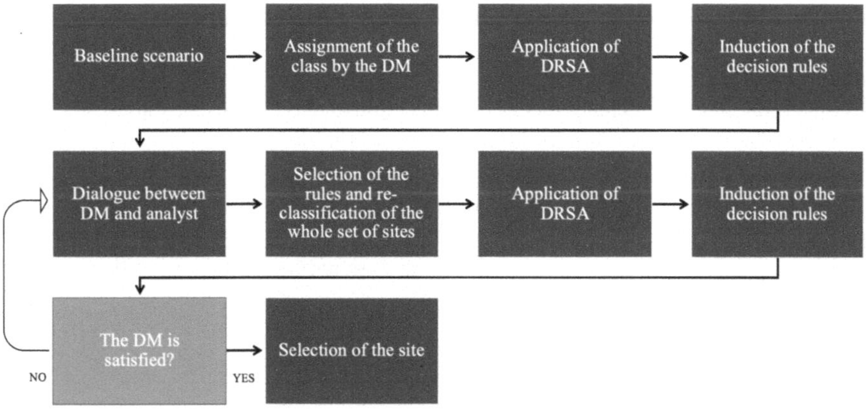

Fig. 6.3 Scheme of the interactive DRSA

4.2.3 Results and Role Played by the MCDA

The reduction from 39 sites to 2 satisfactory solutions is a remarkable achievement for this kind of decision problem. This is particularly true since the selection process is based on arguments supported by a mathematical model allowing the DM to increase the awareness of the problem and to be able to justify the selection with strong arguments that can be clearly communicated.

The interactive DRSA proved to be very useful for the DM in order to reason about the data involved to elicitate the DM's preferences. The "*if*... *then*" decision rules give argumentation in a logical form "speaking the same language" of the DM without any recourse to technical terms. Moreover, the mathematically robust decision rules are able to strongly support the DM in solving the bounded rationality, managing complex problem characterized by different options and qualitative/quantitative attributes.

4.3 Strategic Urban Planning: The MACBETH Method for the Requalification of a District in Brussels Facing Urban and Social Decline

4.3.1 Decision Problem

The case study considered for this application refers to the simulated academic process for the transformation of the Tür und Taxis district in Brussels (Belgium) (Abastante and Lami 2018), subject to a discussion from the Municipality of Brussels. Tür und Taxis is located in the city centre, near the sadly known (for terrorist reasons) Molenbeek district. Despite they are normative separated, they are considered as a big district in the hypotheses of new masterplans of the Brussels municipality. In this sense, the aforementioned districts are social mix collectors characterized by the presence of many bottom-up social initiatives in which different architectural and social characters exist. This contributed causing physical, architectural and social discomfort highlighting of the second immigrant generations living in the districts. In particular, the decision problem conducted was devoted to identify the most interesting criteria to reduce the physical discomfort improving the affordability of houses and the quality of the buildings, constructing massive multifunctional buildings characterized by an external hard shell and a flexible use of internal spaces.

4.3.2 MCDA Method

The MACBETH (Measuring Attractiveness by a Categorical-Based Evaluation Technique) (Bana e Costa et al. 2010) method is a multicriteria analysis based on the additive value model, requiring qualitative judgements about differences of value to help a group of stakeholders quantify the relative attractiveness of alternatives. The MACBETH method can be divided into three main phases: (i) the *Model Structuring* phase, in which alternatives and values of concerns are defined; (ii) the *Evaluating* phase, involving a series of pairwise comparisons, where the stakeholder is asked to express his preferences according to the following semantic categories, Extreme, Very strong, Strong, Moderate, Weak, Very Weak and No (no differences between the elements); and (iii) the *Analysis of the Results* devoted to discuss the results in the form of ranking.

For the case study in exam, two sets of criteria have been defined with the aim of considering both the urban design and the architectural characters of the buildings. The main criteria (economic value, buildings' design, envelop performance and accessibility) have been in turn detailed up to 12 sub-criteria. The criteria and sub-criteria have been organized according to a value tree approach provided by the method (Fig. 6.4).

4.3.3 Results and Role Played by the MCDA

The application of the MACBETH method allowed to identify a priority list among both criteria and sub-criteria to be considered to support the decision process for the case study in exam (Abastante and Lami 2018). The results showed that the buildings' design and the economic values of the transformation are the most important criteria to be considered. In particular, according to the sub-criteria identified, a proper design of the ground floor functions would be fundamental in order to propose a sensible architectural project able to contribute reducing the physical and social discomfort of the Tür und Taxis area. In this sense, the MACBETH method turned out to be useful to simultaneously consider the two levels of the problem: the socio−/territorial and the architectural one.

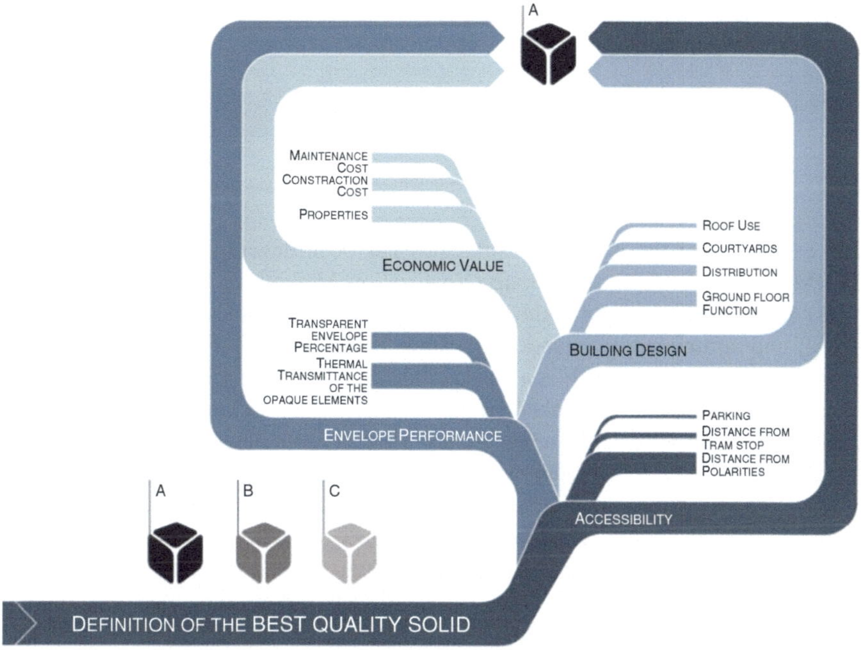

Fig. 6.4 The value tree of the decision problem. (Source: Abastante and Lami 2018)

4.4 Energy Planning: An Application of the PROMETHEE Method for the Definition of a Net-Zero Energy District (nZED)

4.4.1 Decision Problem

As it is well-known, the residential building sector is one of the biggest consumers of energy in the European Union (EU), and it is responsible for the 40% of the CO_2 emissions. In order to face this problem, the European Commission is promoting the use of renewable energy sources, introducing the new standards of nZEB (nearly Zero Energy Building) and nZED (net-Zero Energy District). In particular, the shifting of the attention from the building scale to the urban scale requires to include in the analysis not only the aspects strictly related to energy and environmental impacts but also those concerned with social and economic sectors, such as the number of jobs created by the energy investments, the effects in terms of quality of life of the people involved and so on. The present evaluation considers the energy requalification of a residential district in the municipality of Turin (Northern Italy) (Becchio et al. 2017). The district is constituted by high-rise apartment buildings with different for typology and use and characterized by low thermal properties.

Four alternative strategies are considered for the requalification of the district, which combine different building envelope improvements with specific energy efficiency measures.

4.4.2 MCDA Method

The evaluation of the alternative strategies has been developed by means of the PROMETHEE (Preference Ranking Organization Method for Enrichment Evaluations) method, which is one of the most known MCDA methods drawn up by Brans (1982) and subsequently extended by Brans et al. (1984, 1986) and Vincke and Brans (1985). PROMETHEE is a non-parametric outranking method for a finite set of alternative actions to be ranked on the basis of different criteria, which are often conflicting. Each alternative is valued by a positive or negative preference flow through a value outranking relation, in order to determine how much an alternative is outranked compared to the others.

A family of qualitative/quantitative criteria has been defined for the evaluation of the alternative requalification strategies, including not-renewable energy consumption, environmental emissions production, global cost of the operation, indoor comfort for the inhabitants, green jobs creation and real estate market value increase. Following the PROMETHEE methodology, a preference function has been defined for each criterion which allows to establish how much an alternative is preferred over another. The second type of information which characterizes the evaluation model regards the preference weights assumed by the decision-makers about a specific criterion, to determine the relative importance of one over another. To this purpose, personal interviews with experts in different fields were carried out through the set of cards methodology (Figueira and Roy 2002) (Fig. 6.5).

4.4.3 Results and Role Played by the Evaluation

The PROMETHEE result aggregation (Fig. 6.6) indicates that strategy 1 and strategy 2, characterized by biomass energy carrier, are generally preferred over the alternative with fossil fuel carrier by all experts, highlighting the interest for environmental impacts. Strategy 2, thanks to its higher comfort level, achieves a higher satisfaction compare to strategy 1. On the other hand, strategy 3 turned out to be the one with the worse performance according to all experts. The results show the advantages of using the PROMETHEE method for this particular problem. In particular the PROMETHEE II complete ranking allows to overcome the problem of the incomparability between actions. This is a very important strength of the method because it provides the DM with a clear ranking that is useful to base the final decision.

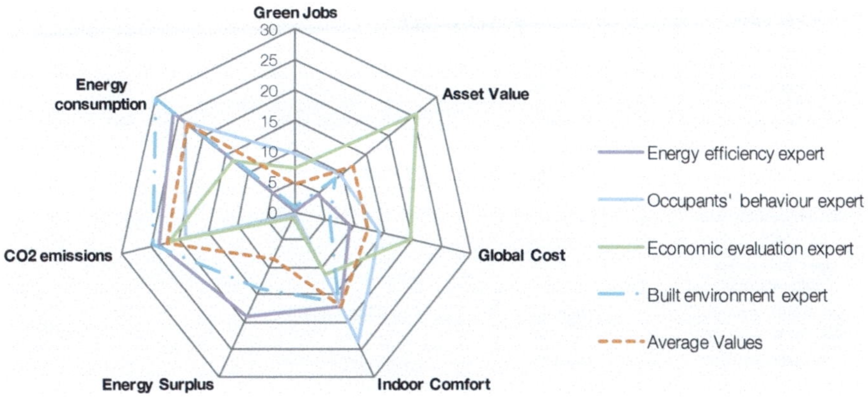

Fig. 6.5 Sets of weights resulting from the different experts. (Source: Becchio et al. 2017)

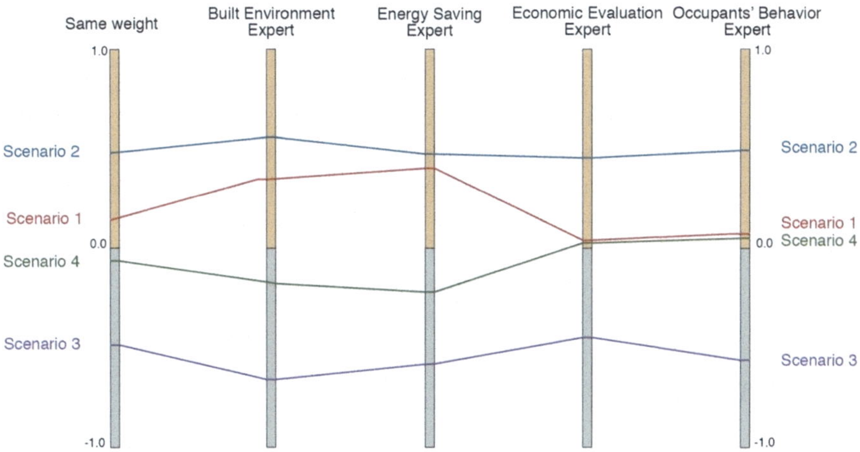

Fig. 6.6 Ranking comparison for the different experts. (Source: Becchio et al. 2017)

4.5 Cultural Adaptive Reuse of Abandoned Buildings

4.5.1 Decision Problem

In this case study, a decision-maker (DM) faces a decision situation with respect to the adaptive reuse for cultural and art purposes of underused sites and buildings located in Turin, Italy (Costa et al. 2019). As other Italian and European cities, Turin has faced with the double problem of the deep economic crisis and the change of the urban fabric, with ten million square meters of underused industrial areas until the end of the 1990s. From those years, one of the elements characterizing the strategy of

CRITERIA	SCALE LEVELS		DIRECTION
Temperature (Temp)	Cardinal	0 – 7,5	Min
Humidity difference (RH)	Cardinal	0 - 12,5	Min
Overshadow possibility (O)	Ordinal	Low, Medium, High	Max
Sound quality (SQ)	Ordinal	Low, Medium, High	Max
Site surface difference (SS)	Cardinal	0 - 9925	Min
Open space surface difference (SS)	Cardinal	0 - 35	Min
Suitability to improve technology service (Tech)	Ordinal	Low, Medium, High	Max
Walkability (W)	Ordinal	Low, Medium, High	Max
Distance from public transport (DistPT)	Cardinal	0 - 3500	Min

Fig. 6.7 Criteria and their scale. (Source: Costa et al. 2018)

repositioning the city of Turin was the decision to put culture—intended primarily as an identity element of the territory—at the centre of its transformation and internationalization process. In this perspective, a set of 15 places was considered, including underused buildings and abandoned open spaces characterized by a current state of disuse, but with different surfaces and conditions. They included former factories, churches, villas, public buildings, entertainment buildings and open-air spaces. In this context, the DM wanted to define the most appropriate cultural destination (among six categories defined) for the fifteen places considered.

4.5.2 MCDA Method

The problem at hand was a nominal classification problem, in which the categories are not preferentially ordered. In fact, no preference order exists among the kind of events to each building to be reused can be assigned. The aim was to know which are the most appropriate events to each building taking into account its specific characteristics. The CAT-SD (CATegorization by Similarity-Dissimilarity) method has been used, based on similarity and dissimilarity concepts, in order to help a DM to deal with multiple criteria nominal classification problems (Costa et al. 2018). This method was designed for aiding the DM to classify actions (or alternatives) into not preferentially ordered categories, considering a way to model similarity and dissimilarity between two actions. The method considers several criteria to assess actions (Fig. 6.7) and uses a set of reference actions to define the categories. In this method, distinct sets of criteria weights, interaction effects between some pairs of criteria and likeness thresholds can be considered per category. The classification of an action to a category depends on its comparison to the set of reference actions, resulting in a likeness degree, and the likeness threshold of the category. In some cases, for the elicitation of preference parameters to construct a CAT-SD decision model, it is recommended to follow a constructive approach involving the analyst (or facilitator) and the DM (i.e. they interact during the decision process in a co-constructive way) (Roy 1993). As a result, transparency can be added to decision aiding process in the modelling phases.

4.5.3 Results and Role Played by the MCDA

The CAT-SD method seems to be extremely suitable as a support tool for a DM towards a decision related to the adaptive reuse, where the potential of an abandoned space is central to program definition. It could help in fact to reason on the versatility of the places and suitability for specific uses, as well as to build an appropriate vision that links the existing infrastructure and its potential to local and global trends. The specific case study in Costa et al. (2019) has moreover the wish to "model" an elusive theme such as the artistic one, presenting numerous obvious difficulties. Features like the surface of the area, the temperature and the presence of open spaces can vary for every cultural event in relation to different aspects, such as the will of the artist or the particularity of the event itself. In this sense, every cultural event represents a unique case: the consequence is that the definition of the general paradigms to be used as input data in the computations has been a considerable challenge.

More in general, the CAT-SD method can be applied in any decision related to territorial and urban planning with the aim of supporting stakeholders and experts to define sustainable destination of areas and buildings of interests through an adequate multiple criteria decision aiding-based nominal classification procedure.

4.6 Built Environment: Application of the NAROR for the Management of a Real Estate Portfolio

4.6.1 Decision Problem

This case study concerns the investigation of a real estate portfolio. In particular, in 2007 the city of Turin has been the first Italian city to have promoted the constitution of a real estate investment trust for selling some of its unused properties. The fund was constituted by 18 estates, for a total surface of more than 80,000 m^2 (Fig. 6.8). The properties are mainly situated in the city centre and in the hillside, so in areas characterized by a high medium value of the estates. The objective of the evaluation is related to the choice of the best performing property in terms of valorization on the real estate market (Bottero et al. 2016).

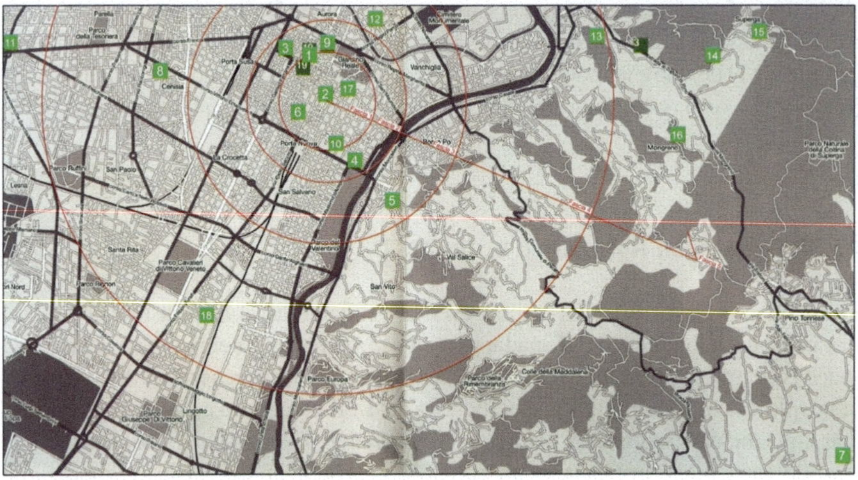

Fig. 6.8 Location of the considered real estate properties

4.6.2 MCDA Method

In the evaluation, the Non-Additive Robust Ordinal Regression (NAROR) has been applied (Angilella et al. 2010a, b, 2016). NAROR is a recent extension of the Robust Ordinal Regression family of MCDA methods (Greco et al. 2008b; Figueira et al. 2010; Corrente et al. 2013, 2014) to the Choquet integral preference model which permits to represent interaction between considered criteria through the use of a set of non-additive weights called capacity or fuzzy measure.

For evaluating the alternatives under examination, a family of six criteria has been constructed which includes location of the building, current destination, age, flexibility of the estate to the transformation, estimated value of the asset and surface. A panel of experts has been created for simulating the decision-making process. The panelists provided preference information in terms of preferences between pairs of alternatives and relative importance and interaction of considered criteria.

Figure 6.9 represents the exploitation of the preference model obtained from the work with the experts' panel.

Even if this information is very meaningful, in general the DM would like to have one ranking of all the considered sites. For this reason, the most representative value function (MRVF) has been computed considering the results of the NAROR and giving a total order of all the assets at hand.

4.6.3 Results and Role Played by the MCDA

According to the results of the calculation, the asset 2 turned out to be the best performing solution. In the lights of these results, the arguments at disposal of the

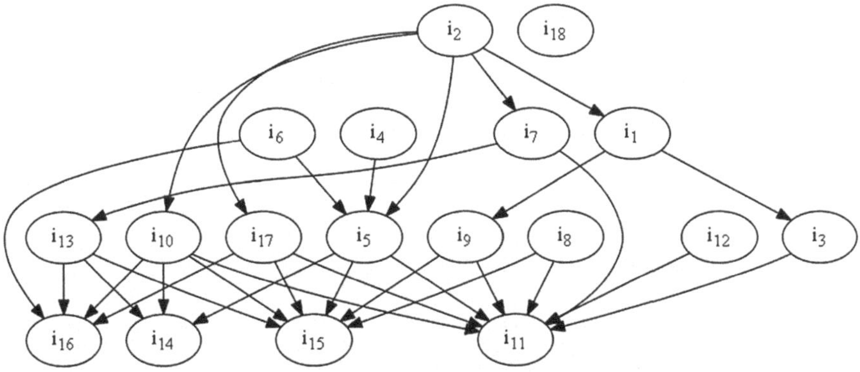

Fig. 6.9 Graphical representation of the necessary preference relations

DM for selecting the asset 2 for the starting of operation are solid enough for justifying the choice.

The application of the NAROR to the case shows that the method has numerous advantages in the context of complex territorial decision problems. Firstly, the method allows to manage a quite high number of alternatives by means of a limited number of preference information required, providing very clear and readable elements, even if the mathematical procedure is rich and articulated. Secondly, the method is suitable for developing an integrated assessment able to consider the interaction between criteria, which is a more and more an interesting research issue in sustainability assessments.

4.7 *Environmental Systems: An Extension of the ELECTRE III Method with Interactions for the Evaluation of Requalification Scenarios*

4.7.1 Decision Problem

The decision problem under analysis concerns the requalification of an abandoned quarry located in the Province of Novara (Northern Italy) (Bottero et al. 2015). The quarry has been abandoned since 1975 and covers a total surface of 65,000 m2. Due to its abandoned state, the area is now suffering from uncontrolled vegetation growth and water-filled pits. Despite these critical conditions, the area has a high potential as it is part of the provincial ecological system of environmentally valuable sites. For the reclamation of the area, five alternative projects are considered by the Municipal Authority that can be described as follows: (1) basic reclamation, (2) realization of a forest, (3) development of a wetland, (4) implementation of the ecological network and (5) construction of a recreational structure.

	1. Cost	2. Profitability	3. Services	4. Landscape ecology	5. Environment	6. Planning
1. Cost					$k_{15} = 14$	
2. Profitability						
3. Services						
4. Landscape ecology		$k'_{24} = 6$			$k_{45} = -9$	
5. Environment	$k_{51} = 14$	$k'_{25} = 6$		$k_{54} = -9$		
6. Planning						

Fig. 6.10 Interaction coefficients between criteria defined by the experts (Bottero et al. 2015)

4.7.2 MCDA Method

An extension of the ELECTRE III method for taking into account criteria interactions has been applied for the problem under examination (Figueira et al. 2009). In particular, three types of interaction are considered in the evaluation: mutual strengthening, mutual weakening and antagonistic effect. A mutual strengthening (or synergy) between two criteria occurs when the importance of the criteria together is greater than the sum of the importance of the two criteria considered separately; conversely, a mutual weakening (or redundancy) between two criteria occurs when the importance of the criteria together is smaller than the sum of the importance of the two criteria considered separately; finally, the antagonistic effect is able to model the situations in which two criteria are strongly divergent.

Starting from the overall objective of the analysis, which is the identification of the most sustainable project for the reuse of the abandoned quarry, a coherent set or family of criteria that reflects all the concerns relevant to the decision problem has been identified, considering the relevant literature in the field of environmental planning and the requirements coming from the legislative framework in the context of Environmental Impact Assessment. The model is based on qualitative and quantitative criteria which include investment cost, profitability, new services for the population, landscape ecology effects, environmental impacts and consistency with local planning requirements. A focus group of experts (in economic evaluation, environmental engineering and landscape ecology) was constituted for the assignment of numerical values to the weights and to the interaction coefficients (Fig. 6.10).

4.7.3 Results and Role Played by the MCDA

ELECTRE III with interactions between criteria leads to rank strategy 2 (forest) as the most sustainable solution for the requalification of the area, followed by strategy 3 (wetland). The results of the evaluation model have been verified by means of specific sensitivity and robustness analysis which demonstrated that this ranking is valid for whatever the considered sets of weights and interaction coefficients.

Apart from the coherence of the obtained results, the research seems to have a particular interest in the context of the methods for sustainability assessment. In fact, when dealing with sustainability assessment in an integrated way, a critical issue is how to combine the different dimensions in the evaluation framework. In this context, of particular importance is the degree of compensability between the different dimensions/aspects of the problem as it has been noticed that neither an economic reductionism nor an ecological one is possible (Munda 2005). Since in general, economic sustainability has an ecological cost and ecological sustainability has an economic cost, an integrative evaluation framework like the one here presented is needed for tackling sustainability issues properly.

5 Lessons Learned for Sustainability Assessment

In the context of territorial transformations, where the decisional arena is characterized by a plurality of stakeholders with different values and objectives, Multiple Criteria Decision Analysis is a valuable tool for the integrated assessment of alternative scenarios, being able to consider simultaneously both technical elements, which are based on empirical data and observations, and non-technical elements, which are based on social visions, preferences and feelings (see for examples the case studies presented in Sects. 6.4.3 and 6.4.5).

One of the main advantages of the approach is related to the possibility of including in the evaluation model the opinion of relevant actors and stakeholders by means of different participatory techniques such as interviews, questionnaires, focus groups or workshops. For example, in the case study of Sect. 6.4.1, a specific collaborative workshop has been organized for discussing with the stakeholders the structuring of the decision problem and the importance of the different aspects; the application presented in Sect. 6.4.7 is grounded on the creation of a multidisciplinary experts' panel for the definition of the parameters to be used in the model. Under this perspective, the evaluation is not seen as "one-shot activity", rather as a social learning process where DMs and stakeholders learn about the problems under examination, while they are solving them. In this sense, MCDA allows the creation of a common knowledge among DMs, local communities and final users, thus ensuring the increase in the social capital.

Another strength of MCDA approach is related to the possibility for the DMs to represent in a clear way their preferences, providing strong arguments for the justification of the final choice. This aspect is particular clear in the application of the DRSA method presented in Sect. 6.4.2 and in the development of the NAROR technique illustrated in Sect. 6.4.6.

It is also worth mentioning that, despite the coherence of the results provided by a MCDA model, the findings of the evaluation can be further verified by means of specific sensitivity analysis that allow to test under which conditions the obtained results are confirmed (see the example of Sect. 6.4.4).

MCDA offer also flexible frameworks for the communication of the results of the evaluation process, making use of diagrams, charts and tables. Of particular importance is the possibility of integrating MCDA model with GIS tools that allow the production of spatial maps (see the example of Sect. 6.4.1) able to clarify the complexity of the reality in order to provide a rational basis for understanding complex information and to create a common knowledge platform.

6 Conclusions

The chapter illustrates a selection of methodologies applied to seven territorial and urban transformation projects, where the "sustainability question" challenges the experts to provide the most appropriate methodology to support the DMs. The case studies are real-world applications in Europe, with a scale that varies from the urban level to the transnational one, and a large variety of realms (transport, undesirable facilities, urban planning, energy retrofitting, real estate investments, cultural adaptive reuse and environmental systems). This variety allowed us to synthetically discuss the possible contribution to the decision-making process of the different MCDAs: it can be stated that all are effective (with a relevance that goes from medium to extreme), but some are more suitable than others based on the nature of the problem, the data available and the type of support required in the decision. Effectiveness is also demonstrated by the fact that the MCDAs are able to tackle the ever-wider concept of sustainability, now transversal in every decision, and present as "the issue" in the political agendas of the whole planet.

References

Abastante, F., & Lami, I. M. (2013). An analytical model to evaluate a large scale urban design competition. *Geoingegneria Ambientale Mineraria (GEAM), 139*, 27–36.

Abastante, F., & Lami, I. M. (2018). An integrated assessment framework for the requalification of districts facing urban and social decline. *Green Energy and Technology*, 535–545. https://doi.org/10.1007/978-3-319-78271-3_42.

Abastante, F., Bottero, M., Greco, S., & Lami, I. M. (2012). A dominance-based rough set approach model for selecting the location for a municipal solid waste plant. *Geoingegneria Ambientale Mineraria (GEAM), 137*, 43–54.

Abastante, F., Gunther, F., Lami, I. M., Masala, E., Pensa, S., & Tosoni, I. (2014). Analytic network process, interactive maps and strategic assessment: The evaluation of Corridor24 alternative development strategies. In I. M. Lami (Ed.), *Analytical decision-making methods for evaluating sustainable transport in European corridors* (pp. 205–232). Cham: Springer International Publishing AG.

Angilella, S., Greco, S., & Matarazzo, B. (2010a). Non-additive robust ordinal regression: A multiple criteria decision model based on the Choquet integral. *European Journal of Operational Research, 201*(1), 277–288.

Angilella, S., Greco, S., & Matarazzo, B. (2010b). The most representative utility function for non-additive robust ordinal regression. In E. Hullermeier, R. Kruse, & F. Hoffmann (Eds.), *Proceedings of IPMU 2010, IPMU 2010 (220–257), LNAI 6178*. Heidelberg: Springer.

Angilella, S., Bottero, M., Corrente, S., Ferretti, V., Greco, S., & Lami, I. M. (2016). Non-additive robust ordinal regression for urban and territorial planning: An application for siting an urban waste landfill. *Annals of Operations Research, 245*(1–2), 427–456.

Bana e Costa, C. A., & Chargas, M. P. (2004). A career choice problem: An example of how to use MACBETH to build a quantitative value model based on qualitative value judgments. *European Journal of Operational Research, 153*, 323–333.

Bana e Costa, C. A., De Corte, J. M., & Vansnick, J. C. (2010). MACBETH: Measuring attractiveness by a categorical based evaluation technique. In J. J. Cochran (Ed.), *Encyclopedia of operations research and management science*. New York: Wiley.

Barbier, E. (1987). The concept of sustainable economic development. *Environmental Conservation, 14*(2), 101–110.

Becchio, C., Bottero, M., Corgnati, S., & Dell'Anna, F. (2017). A MCDA-based approach for evaluating alternative requalification strategies for a Net-Zero Energy District (NZED). In C. Zopounidis & M. Doumpos (Eds.), *Multiple criteria decision making. Application in energy and management* (pp. 189–211). Berlin: Springer.

Bottero, M., & Lami, I. M. (2010). Analytic network process and sustainable mobility: An application for the assessment of different scenarios. *Journal of Urbanism, 3*, 275–293.

Bottero, M., Ferretti, V., Figueira, J. R., Greco, S., & Roy, B. (2015). Dealing with a multiple criteria environmental problem with interaction effects between criteria through an extension of the Electre III method. *European Journal of Operational Research, 245*(3), 837–850.

Bottero, M., Angilella, S., Corrente, S., Ferretti, V., Greco, S., & Lami, I. M. (2016). *An application of the NAROR for the management of a real estate portfolio*. European conference on operational research EURO 2016 (Poznan, 3–6 July 2016).

Brans, J. P. (1982). *L'ingénierie de la décision. Elaboration d'instruments d'aide à la décision. Methode PROMETHEE. Instruments et Perspectives Davenir* (pp. 183–214). Quebec: Universite Laval.

Brans, J. P., Mareschal, B., & Vincke, P. (1984). PROMETHEE: A new family of outranking methods in multicriteria analysis. *Operational Research, 84*, 447–490.

Brans, J. P., Mareschal, B., & Vincke, P. (1986). How to select and how to rank projects: The PROMETHEE method. *European Journal of Operational Research, 24*(2), 228–238.

Bruntland, G. (Ed.). (1987). *Our common future*. Oxford: Oxford University Press.

Corrente, S., Greco, S., Kadzinski, M., & Słowi'nski, R. (2013). Robust ordinal regression in preference learning and ranking. *Machine Learning, 93*(2–3), 381–422.

Corrente, S., Greco, S., Kadzinski, M., Słowi'nski, R. (2014). Robust ordinal regression. *Wiley Encyclopedia of Operations Research and Management Science*, (pp. 1–10). Hoboken, US: Wiley

Costa, A. S., Figueira, J. R., & Borbinha, J. (2018). A multiple criteria nominal classication method based on the concepts of similarity and dissimilarity. *European Journal of Operational Research, 271*(1), 193–209.

Costa, A. S., Lami, I. M., Greco, S., Figueira, J. R., & Borbinha, J. (2019). A multiple criteria approach dening cultural adaptive reuse of abandoned buildings. In M. H. S. Geiger & A. de Almeida (Eds.), *Multiple criteria decision-making and aiding – Cases on decision-making methods and models with computer implementations* (pp. 193–218). Cham, Switzerland: Springer.

De Magalhaes, C. (2015). Urban Regeneration. In *International encyclopedia of the social & behavioral sciences*. Amsterdam: Elsevier.

Faludi, A. (Ed.). (1973). *A reader in planning theory* (Urban and regional planning series 5). New York: Pergamon Press.

Figueira, J., & Roy, B. (2002). Determining the weights of criteria in the ELECTRE type methods with a revised Simos' procedure. *European Journal of Operational Research, 139*, 317–326.

Figueira, J., Greco, S., & Ehrgott, M. (Eds.). (2005). *Multiple criteria decision analysis. State of the art survey*. New York: Springer.

Figueira, J., Greco, S., & Roy, B. (2009). Electre methods with interaction between criteria: An extension of the concordance index. *European Journal of Operational Research, 199*, 478–495.

Figueira, J., Greco, S., & Słowi'nski, R. (2010). Building a set of additive value functions representing a reference preorder and intensities of preference: GRIP method. *European Journal of Operational Research, 195*(2), 460–486.

Greco, S., Matarazzo, B., & Słowiński, R. (2001). Rough sets theory for multicriteria decision analysis. *European Journal of Operational Research, 129*, 1–47.

Greco, S., Matarazzo, B., & Słowiński, R. (2008a). Dominance-based rough set approach to interactive multiobjective optimization. In J. Branke, K. Deb, K. Miettinen, & R. Słowiński (Eds.), *Multiobjective optimization, LNCS* (Vol. 5252, pp. 121–155). Heidelberg/New York: Springer-Verlag Berlin.

Greco, S., Mousseau, V., & Słowinski, R. (2008b). Ordinal regression revisited: Multiple criteria ranking using a set of additive value functions. *European Journal of Operational Research, 191* (2), 416–436.

Healey, P. (2007). *Urban complexity and spatial strategies. Towards a relational planning for our times*. London/New York: Routledge.

Lami, I. M. (2014). Evaluation tools to support decision-making process related to European corridors. In I. M. Lami (Ed.), *Analytical decision-making methods for evaluating sustainable transport in European corridors* (pp. 85–102). Cham: Springer International Publishing AG.

Lami, I. M. (2019). The context of urban renewals as a 'Super-Wicked' problem. In F. Calabrò, L. Della Spina, & C. Bevilacqua (Eds.), *New metropolitan perspectives. ISHT 2018. Smart innovation, systems and technologies* (Vol. 100, pp. 249–255). Cham: Springer.

Lami, I. M., Abastante, F., Bottero, M., Masala, E., & Pensa, S. (2014). A MCDA and data visualization framework as a problem structuring method (PSM) to address transport projects. *Euro Journal of Decision Processes (EJDP), Special ISSUE on Problem Structuring, 2*(3–4), 281–312.

Mondini, G. (2008). Il progetto di sostenibilità. In P. Lombardi (Ed.), *Riuso edilizio e rigenerazione urbana. Innovazione e partecipazione*. Torino: Celid.

Munda, G. (2005). Social multi-criteria evaluation for urban sustainability policies. *Land Use Policy, 23*, 86–94.

Pensa, S., Masala, E., & Lami, I. M. (2013). Supporting planning processes by the use of dynamic visualization. In S. Geertman, F. Toppen, & J. Stillwell (Eds.), *Planning support Systems for Sustainable Urban Development* (pp. 451–467). Berlin Heidelberg: Springer.

Roy, B. (1993). Decision science or decision-aid science? *European Journal of Operational Research, 66*(2), 184–203.

Roy, B., & Bouyssou, D. (1993). *Aide Multicritére à la Décision: Méthodes et cas*. Paris: Economica.

Roy, B., & Slowinski, R. (2013). Question guiding the choice of a multicriteria decision aiding method. *EURO Journal on Decision Processes, 1*(1), 69 97.

Ruegg, J. (1994). Formes du PPP. In J. Ruegg, S. Decoutère, & N. Mettan (Eds.), *Le partenariat public-privé. Presses Polytechniques et universitaires Romandes* (pp. 79–98). Lausanne.

Saaty, T. L. (2005). *Theory and applications of the analytic network process*. Pittsburgh: RWS Publications.

Secchi, B. (1984). *Il racconto urbanistico*. Torino: Einaudi.
Simon, H. A. (1960). *The new science of management decision*. New York: Harper and Brothers.
sustainabledevelopment.un.org.
Vincke, J. P., & Brans, J. P. (1985). A preference ranking organization method. The PROMETHEE method for MCDM. *Management Science, 31*, 641–656.

Chapter 7
Sustainable Supplier Segmentation: A Practical Procedure

Hamidreza Fallah Lajimi

Abstract Sustainability of supply chain is determined by the sustainability performance of each partner in the chain. The relationship between buyer and supplier has an important role in improving the sustainability of supply chain. The purpose of this chapter is to explain the process of segmenting the sustainable suppliers and presenting strategies for collaboration and improvement of the suppliers. In this chapter, a six-step process is suggested in which the output of one step is considered as the input of the other step. At first, the performance indicators in each dimension of sustainability are introduced, and then given the performance indicators in the segmentation model, the sustainable suppliers are assigned into seven segments (three main segments, economic, social, and environmental; three balancing segments, bearable, viable, and equitable; and supplementary segment, sustainable). Finally, some supplier development strategies appropriate to each dimension of sustainability are suggested. The present study is beneficial for the researchers and companies' executive managers. By a deeper understanding of this process, researchers can benefit from the proposed process for sustainable supplier segmentation, and the corporate executives and experts can have the opportunity of using the supplier collaboration and development strategies in the supply chain.

Keywords Sustainable supply chain · Sustainable supplier · Supplier development

1 Introduction

Until the 1960s, the companies' efforts focused on the economic aspects of business, and after this period, companies' attention was directed toward noneconomic aspects of developmental activities such as social and environmental activities. The issue of

H. F. Lajimi (✉)
Department of Industrial Management, Faculty of Economics and Administrative Sciences, University of Mazandaran, Babolsar, Iran
e-mail: h.fallah@umz.ac.ir

J. Rezaei (ed.), *Strategic Decision Making for Sustainable Management of Industrial Networks*, Greening of Industry Networks Studies 8,
https://doi.org/10.1007/978-3-030-55385-2_7

environmental and social business impacts has led to the introduction of a new concept called "sustainable development" which addresses economic, social, and environmental dimensions, simultaneously (Cowell and Parkinson 2003). Sustainable development aims to meet the needs of the present generation without undermining the needs of the next generation. This concept has gradually entered the supply chain and has led to the emergence of the concept of sustainable supply chain. While there are various definitions of sustainability, Seuring and Müller (2008), in a comprehensive definition, defined sustainable supply chain management, as the management of materials and information flows as well as the collaboration of the organizations across the supply chain with regard to sustainable development goals, i.e., economic, social, and environmental goals arising from customers and stakeholders' needs. Consideration of the three social, environmental, and economic perspectives has been observed in most definitions of sustainability.

Social Perspective Society's expectations of the legal, economic, ethical, and humanitarian activities of organizations indicate their social responsibility (Carroll and Buchholtz 2002). Social responsibility is a tool for sustaining the organization and achieving the sustainable development goals to improve the people's life quality without damaging the environment and with minimizing the use of resources and ensuring the economic rights of citizens (Carter and Rogers 2008). Variety of choices, human rights standards, job creation and creation of self-employment opportunities, and improvement of community health infrastructures are important elements of social sustainability and should be considered in supply chain assessment (Bai and Sarkis 2010; Martínez-Blanco et al. 2014). In recent years, the concept of corporate social responsibility has been considered by researchers and companies in the manufacturing and service industries as a competitive advantage (Govindan et al. 2013).

Environmental Perspective In recent decades, changes in environmental elements have directed the researchers' interest toward developing environmental management strategies in the supply chain (Fernando and Saththasivam 2017). Environmental issues are among the key elements of sustainable development that refer to the conservation of natural resources and the protection of the environment from the dangers of technological progress. Environmental sustainability aims at using the current generation resources properly (without disrupting future generation's use of resources) and considering the environmental requirements at all stages of selection and supply of raw materials, production, distribution, delivery, and recycling management in order to maximize the energy and resource consumption efficiency with emphasis on improving performance (Handfield et al. 1997; Vachon and Klassen 2008; Eltayeb et al. 2011). Considering environmental issues has advantages such as loss reduction and cost savings, creating competitive advantage, increasing environmental performance, and improving synergy and efficiency (Govindan et al. 2013; Zhu et al. 2008).

Economic Perspective Usually in designing supply chains and businesses, economic aspects such as cost reduction have been emphasized upon (Cowell and

Parkinson 2003), and many studies have considered the optimal design of the supply chain from the economic perspective (Chaabane et al. 2012). Economic sustainability is the interaction between social and environmental sustainability factors that are essential for the organization's long-term life and economic growth (Yusuf et al. 2013). The goal of economic sustainability, in addition to increasing profitability, is to perform processes that do not interfere with environmental and social activities (Adetunji et al. 2003). Financial performance, economic crisis management, financing of key economic and community infrastructures, and supporting the long-term economic development in the society are considered as important components in assessing economic sustainability (Wu and Shen 2013).

Large companies adjust their performance criteria based on sustainable supply chain to gain competitive advantage over other competitors. For example, Deutsche Bahn is one of the leading companies in the domain of travel and logistics which is active in 130 countries of the world. In 2018, this company adjusted its business plan based on sustainability and 17 goals for supply chain sustainability development. Moreover, performance criteria were presented to achieve sustainability goals in the economic (punctuality, high) and social dimensions. Performance criteria were also presented to achieve sustainability goals in economic (punctuality, high-performance infrastructure, profitability, digitalization and innovation, financial stability, performance capability, etc.), social (customer satisfaction, employee satisfaction, employer attractiveness, etc.), and environmental (climate protection, noise reduction, etc.) dimensions. Concurrent with adopting sustainability strategies, much research had been conducted in this area as well. Carter and Rogers (2008) introduced the concept of sustainability in supply chain management through the comprehensive review of literature and use of conceptual theory and explored the relationship among environmental, social, and economic performance in the area of supply chain management. They presented a supply chain sustainability framework, based on resource dependence theory, transaction cost economics, population ecology, and the resource-based view of the firm.

2 Importance of Supplier Segmentation and Its Role in Overall Supply Chain Performance

Supplier relationship management involves activities related to buyer-producer interaction. In recent years, many researchers investigated different areas of supplier relationship management, and it has not yet been investigated as a whole (Glock et al. 2017). Evaluating suppliers in the supply chain is one of the crucial and vital elements in the overall performance of the chain and in improving the competitiveness of the supply chain which is essential for establishing a sustainable relationship between the manufacturer and customer needs (Xu et al. 2019; Jin et al. 2013). Important processes in supplier evaluation include selection, segmentation, monitoring, and control of suppliers that are qualified (Segura and Maroto 2017). Supplier segmentation is a strategic process in supplier management that falls between two processes of selection and supplier relationship management (Rezaei and Ortt 2012). The role and importance of supplier segmentation in supplier relationship management is very significant, and failure to properly execute it by the buyer company can result in wasted time, too much cost, and collaborating with undesirable suppliers. Rezaei and Ortt (2012) stated that:

Supplier segmentation is referred to as grouping a number of suppliers that share common and similar characteristics and formally it is defined as follows (Rezaei and Ortt 2012):

> Supplier segmentation is the identification of the capabilities and willingness of suppliers by a particular buyer in order for the buyer to engage in a strategic and effective partnership with the suppliers with regard to a set of evolving business functions and activities in the supply chain management

Supplier segmentation was examined by a number of researchers using a variety of approaches and techniques. Parasuraman (1980) was the first to discuss the issue of supplier segmentation. In his descriptive study, he introduced a four-step process for supplier segmentation, which was based on customer segmentation. He argued that the criteria used in supplier segmentation should be based on customer-related criteria. Kraljic (1983) can be considered as the most influential scholar in the area of supplier segmentation. The Kraljic matrix, based on supply risk and profit impact, divides goods into four categories: noncritical items, bottleneck items, leverage items, and strategic items. He proposed strategies for dealing with suppliers in each of these sectors. Following Kraljic, many researchers have applied this approach. As mentioned earlier, Kraljic's approach was based on product characteristics and did not include SRM. In their study, Olsen and Ellram (1997) proposed a three-step portfolio model based on different kinds of supplier relationships: (Step 1) analyzing the company's purchases, (Step 2) analyzing the supplier relationships, and (Step 3) developing action plans. They also proposed a two-dimensional approach to supplier segmentation based on the difficulty of managing the purchase situation and strategic importance of the purchase – along with its criteria). Svensson (2004) conducted his quantitative research on supplier segmentation in the automotive industry. A two-dimensional approach was used based on supplier commitment and the importance of commodities among vehicle manufacturers, and accordingly, four supplier relationship strategies were introduced. Finally, a four-step process designed to manage supplier relationships was proposed: analysis of the business environment, analysis of the relationship criteria, selection of the relationship strategy, and managerial decisions involving the relationship strategy. Day et al. (2010) found that all studies on supplier segmentation, up to that point, were conceptual or based on questionnaires and case studies. The authors conducted a thorough review of the available articles in terms of market conditions, supplier characteristics, buyer characteristics, and buyer and supplier relationships. Rezaei and Ortt (2012) proposed a different approach to supplier segmentation, which became the basis for many other studies. They developed a new approach based on three requirements:

long-term potential-based segmentation, involvement of other business functions beyond purchasing in segmentation process, and considering supplier segmentation as one of the steps in selection process and relationship with suppliers. They introduced a two-dimensional approach called the supplier potential matrix (SPM), involving capability and willingness, along with their criteria. In another study, Rezaei and Ortt (2013) applied a multi-criteria decision-making (MCDM) technique to supplier segmentation, while Rezaei et al. (2015) later used a new method called Best-Worst Method (BWM) to segment the suppliers. In the paper by Bai et al. (2017), which dealt with supplier segmentation in green supply chain, using the SPM approach, the criteria of these two dimensions were identified and used to incorporate clustering techniques and VIKOR to segment the suppliers of a large chemical company. A comprehensive review of the literature on supply chain sustainability (see Martins and Pato (2019)) reveals that supplier segmentation has not yet received enough attention in the context of sustainable supply chain.

3 Sustainable Supplier Segmentation

In the process of supplier relationship management, supplier segmentation is part of this process. Supplier relationship process involves identifying, selecting, segmenting, improving, and evaluating suppliers (Rezaei and Ortt 2012; Glock et al. 2017). These activities cannot be done separately and independently, since they must be carried out continuously, and the output of each step is considered as the input of the next step. Figure 7.1 presents a general sustainable supplier relationship management process (SSRMP).

As can be seen from Fig. 7.1, supplier segmentation cannot be done without selecting, improving, and evaluating the suppliers.

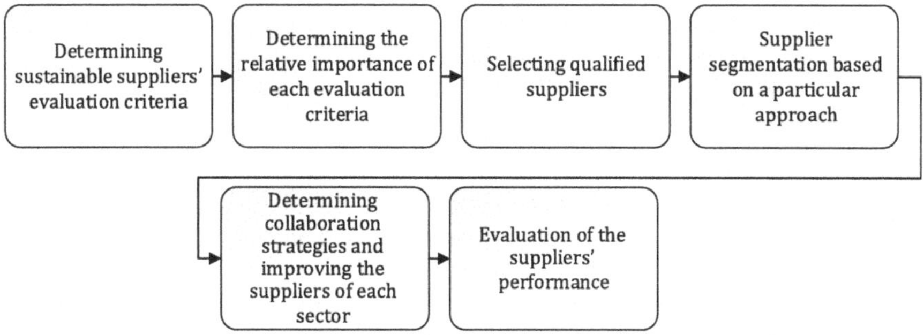

Fig. 7.1 Sustainable supplier relationship management process (SSRMP)

3.1 Determining Sustainable Suppliers' Evaluation Criteria

The first step in evaluating suppliers is selecting appropriate criteria to properly evaluate suppliers. In other words, these criteria should be in line with the goals of the buyer and meet the needs of the customers. Proper and efficient evaluation criteria help reducing risk in supplier selection (Memon et al. 2015) as well as reducing operating costs (Wang et al. 2009). Given that in the sustainable supply chain, simultaneous consideration is given to the three economic, social, and environmental concepts; therefore classification of evaluation indices in these three dimensions takes place easily. Different classifications for supplier selection in the supply chain have been put forward in various studies (Rezaei et al. 2014). This classification has been also done in the supply chain sustainability paradigm. The economic and business dimension criteria have been put forward in eight general cases of cost, quality, time, flexibility, innovation, culture, technology, and communication. The criteria of this dimension of sustainability have been the focus of researchers and managers in traditional supply chain approaches. One of the major concerns of companies in collaborating with suppliers is delivering raw materials at the appropriate time with the desired price and quality which are of the most important issues in most supplier selection studies. In the past, social and environmental dimensions had not played any role in supplier selection, while now increasing awareness of environmental issues has made this factor recognized as a competitive advantage for supply chain players (Diba and Xie 2019). However, some supplier selection studies only emphasized on the environmental aspect (Bai and Sarkis 2010), and others emphasized on both social and environmental aspects. Environmental criteria were included in the categories of pollution control, pollution prevention, environmental management systems, resource consumption, and pollution production. In the social dimension, the criteria fall into the categories of employment, security and health, community concerns, the influence of contractual stakeholders, and other stakeholders. A complete classification of sustainability criteria is presented in Table 7.1.

3.2 Determining the Relative Importance of Each Evaluation Indicator

The sustainable supplier evaluation problem is a multi-criteria decision-making problem, since it comes with different (quantitative and qualitative) types of criteria to evaluate suppliers. In multi-criteria decision-making problems, after determining the criteria, their weights will be determined. Selection of decision-making technique for determining weights depends on factors such as independence and dependence of the criteria, less data, ease of computation, and greater consistency of results. In the literature on supplier evaluation, some techniques have been used to evaluate the weights of sustainability criteria in research. The most important of

Table 7.1 Sustainable criteria for supplier evaluation (Rezaei 2018)

Economic and business criteria	Environmental criteria	Social criteria for supplier selection
Cost	***Pollution controls***	***Employment practices***
Low initial price	Remediation	Disciplinary and security practices
Compliance with cost analysis system	End-of-pipe controls	Employee contracts
Cost reduction activities	***Pollution prevention***	Equity labor sources
Compliance with sectoral price behavior	Product adaptation	Diversity
Quality	Process adaptation	Discrimination
Conformance quality	***Environmental management system***	Flexible working arrangements
Consistent delivery	Establishment of environmental commitment and policy	Job opportunities
Quality philosophy	Identification of environmental aspects	Employment compensation
Prompt response	Planning of environmental objectives	Research and development
Time	Assignment of environmental responsibility	Career development
Delivery speed	Checking and evaluation of environmental activities	***Health and safety***
Product development time	***Resource consumption***	Health and safety incidents
Partnership formation time	Consumption of energy	Health and safety practices
Flexibility	Consumption of raw material	***Local communities' influence***
Product volume changes	Consumption of water	Health
Short set-up time	***Pollution production***	Education
Conflict resolution	Production of polluting agents	Housing
Service capability	Production of toxic products	Service infrastructure
Innovativeness	Production of waste	Mobility infrastructure
New launch of products		Regulatory and public services
New use of technologies		Supporting educational institutions
Culture		Sensory stimuli
Feeling of trust		Security
Management attitude/outlook for the future		Cultural properties
Strategic fit		Economic welfare and growth
Top management compatibility		Social cohesion

(continued)

Table 7.1 (continued)

Economic and business criteria	Environmental criteria	Social criteria for supplier selection
Compatibility among levels and functions		Social pathologies
Suppliers organizational structure and personnel		Grants and donations
Technology		Supporting community projects
Technological compatibility		***Contractual stakeholders' influence***
Assessment of future manufacturing capabilities		Procurement standard
Suppliers speed in development		Partnership screens and standards
Suppliers design capability		Consumers' education
Technical capability		***Other stakeholders' influence***
Current manufacturing facilities/capabilities		Decision influence potential
Relationship		Stakeholder empowerment
Long-term relationship		Collective audience
Relationship closeness		Selected audience
Communication openness		Stakeholder engagement
Reputation for integrity		

these techniques are analytic hierarchy process (AHP) (Luthra et al. 2017), analytic network process (ANP) (Tavana et al. 2017), DEMATEL (Su et al. 2016), and Best-Worst Method (BWM) (Garg and Sharma 2018). After determining the technique, the related questionnaire will be prepared and given to the experts/decision-makers to rate the criteria. After collecting the data, in accordance with the rules governing the technique, weights of indices will be obtained. The importance of each indicator varies according to its role in different industries.

3.3 *Selection of the Qualified Suppliers*

At this stage, it is necessary to select the suppliers that meet the minimum requirements and then evaluate them in the next stage (segmentation). Some companies, empirically and based on the previous operating experience of the suppliers, determine this list of suppliers. But recent approaches to supplier selection indicate that companies such as Deutsche Bahn identify qualified suppliers using scientific methods. At this stage, there is no need for accurate quantitative evaluation of suppliers, and the use of non-compensatory decision-making methods can be effective. Among the non-compensatory methods, the conjunctive satisfying method, by specifying minimums for each indicator, can eliminate the options without these minimums. At this stage, some companies also rank their suppliers. In previous studies, the process of communicating with suppliers generally ended at this stage, and these studies were limited to the ranking and selection of suppliers (Glock et al. 2017). But in practice and in recent approaches to supplier relationships, it has been the focus of researchers and companies.

3.4 Supplier Segmentation Based on a Specific Approach

At this stage, after the qualified suppliers were identified, there is a need for sustainable supplier segmentation. The sustainable supply chain has three main economic, social, and environmental dimensions. The interaction between these three dimensions also creates new economic-social, economic-environmental, and socio-environmental dimensions. According to Carter and Rogers (2008), the main goal of sustainable supply chain management is to achieve the strategic and transparent integration of business processes to achieve social and environmental goals to improve the long-term economic performance of the whole chain. They developed a model based on four factors of risk management, transparency, strategy, and culture for a sustainable supply chain (Fig. 7.2).

Suppliers who pay particular attention to all of these three dimensions consider the demands and needs of the customer with their optimal performance, so that they protect the environment, have concern for community development, and contribute to the economic prosperity. In other words, the companies emphasize on the cooperation with sustainable suppliers that focus on all three social, environmental, and economic dimensions. Companies that implement sustainability dimensions strategically have higher economic performance compared to the companies that only follow the goals of one or two dimensions of the triple dimensions (Carter and Rogers 2008). Based on this model, the suppliers can be segmented from the sustainability perspective. It is clear that any supplier can have a desired performance in one, two, or all three dimensions of sustainability. For example, if a supplier has a desired (high) performance in both economic and social dimensions, it will be considered in the economic-social (equitable) segment. Of course, there are suppliers that are not included in any of the triple dimensions of sustainability in the evaluation and segmentation processes and do not meet the minimum requirements for cooperation; however, these suppliers are eliminated at the earlier stage. At this stage, the suppliers need to be evaluated separately and based on the desired criteria for each sustainability dimension. The criterion for high or low performance of suppliers in each dimension is their scores. In this respect, for the scores below the threshold, the performance is considered low, and for scores above the threshold, the performance is considered high. Table 7.2 lists the supplier characteristics of each segment.

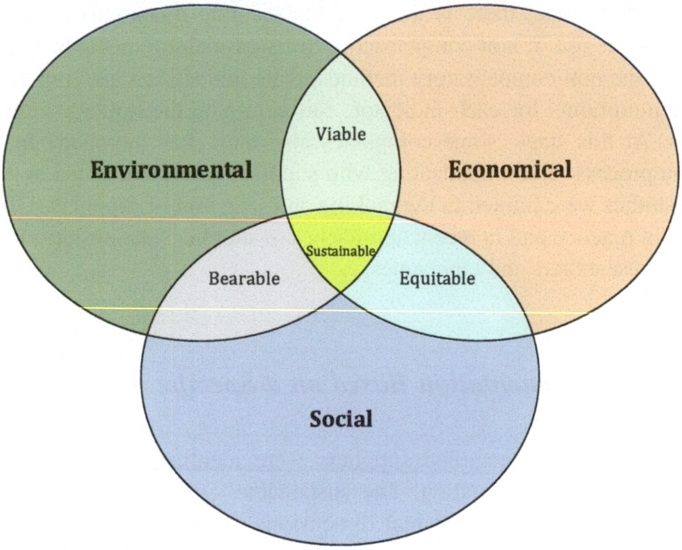

Fig. 7.2 Sustainable supplier segmentation model

3.5 Sustainable Supplier Segmentation Strategies

By reviewing the existing literature related to sustainability, it can be concluded that the supplier improvement program is an issue which has been overlooked in the area of research and even ignored by the companies (Zimmer et al. 2016). However, in order to implement sustainability practices related to suppliers, companies must adopt strategies to collaborate, support, and improve suppliers, since these collaborations facilitate the adoption of innovative environmental technologies (Zhu et al. 2007), adopting sustainability practices (Vachon and Mao 2008), communication, trust, and long-term commitment (Cheung and Rowlinson 2011), adopting ethical procedures in business, improving quality (Talluri et al. 2010), and gaining competitive advantage in marketing and profitability for both the buyer and the supplier (Holt and Ghobadian 2009). In other words, improving suppliers' abilities is needed to increase their level of willingness for long-term cooperation (Rezaei et al. 2015). According to the proposed sustainable supplier segmentation model in this study, companies can consider different collaborative strategies for the suppliers according to the circumstances (for instance, type of industry, type of market, economic conditions). Strategies for collaboration with suppliers should be tailored to the dimensions of industrial marketing such as resource sharing, interdependence, long-term collaboration prospect, collaboration, risk sharing, and supplier improvement (Hadjikhani and LaPlaca 2013). The most important strategies for collaboration and improvement of sustainable suppliers derived from the literature are presented in Table 7.3.

Table 7.2 Characteristics of each segment in the sustainable suppliers segmentation model

Segment	Description	The performance of each dimension		
		Economic	Social	Environmental
Economic	Suppliers of the economic segment focus on creating a profit-driven supply chain and show no interest in other social and environmental aspects. These suppliers emphasize on long-term economic sustainability through increasing market share and value-added profits rather than short-term profits (Björklund et al. 2012), creating new markets and products (Kannan and Tan 2002), improving quality (Wang et al. 2009, 2015), reducing costs through improved efficiency, reducing energy and raw materials consumption (Faisal et al. 2017)	High	Low	Low
Social	Suppliers in social segment focus on the promotion of social justice. The actions taken by these suppliers emphasize on diversity in the workforce, human rights requirements, and nondiscrimination of the employees (ethnicity, gender, religion, age, and social background) and employees' health and safety that are based on the standard quality of life (Faisal et al. 2017; Marshall et al. 2015)	Low	High	Low
Environmental	Environmental suppliers focus on the efficient use of natural resources and the reduction of waste, pollution, and greenhouse gas emissions. Moreover, these suppliers emphasize on the use of renewable raw materials, the elimination of toxic substances, the reduction of energy consumption, and the elimination of destructive waste, with the goal of reducing the negative impacts of raw materials on people's health (Bai and Sarkis 2010; Ortas et al. 2014)	Low	Low	High
Economic-environmental (viable)	In the economic-environmental segment, suppliers with environmental achievement goals consider the economic aspect. Unnecessary costs reduce through the efficient use of energy and natural resources. These suppliers also spend part of their income on providing creative solutions to achieve economic environmental goals. Suppliers of this segment are also called viable	High	Low	High

(continued)

Table 7.2 (continued)

Segment	Description	The performance of each dimension		
		Economic	Social	Environmental
Economic-social (equitable)	Suppliers of socioeconomic segment, which are also referred to as equitable suppliers, focus on supplying raw materials with minimum cost, considering the social needs of society by promoting business ethics, ensuring fair trade and business, and protecting employees' and customers' rights	High	High	Low
Social-environmental (bearable)	In the socio-environmental segment, the emphasis is on having a fair share of natural resources at the national and international levels. Suppliers in this segment pay more attention to the non-economic aspects of business, and to them, people and the environment have higher priority, for which they are also called bearable. From sustainability perspective, paying attention to these suppliers without considering financial and economic outlook would be an irresponsible act (Carter and Rogers 2008), since focusing on environmental and social programs entails some costs	Low	High	High
Sustainable	Sustainable suppliers improve the quality of the end product, thereby enhance people's quality of life, enhance customer satisfaction, provide flexible solutions for future changes, support desired natural and social environments, and lead to the maximum use of the resources with the least costs (Carter and Rogers 2008)	High	High	High

3.6 *Performance Evaluation of Sustainable Suppliers*

In the last phase of supplier segmentation, the supplier performance is evaluated from sustainability perspective. The results of this stage can serve as a feedback for previous stages of sustainable supplier segmentation. Supplier evaluation is an independent process in relation to previous processes of sustainable suppliers' relationship management. This evaluation, which is performed after supplier segmentation, serves as a basis for alternative suppliers, a stimulator of supplier development and improvement activities, as well as a tool for continuous monitoring of the progress of supplier improvement plans. At this stage, supplier efficiency can be determined from sustainability perspective by using mathematical models such as DEA, and consequently the inefficient suppliers can also be determined. As stated

Table 7.3 Strategies for sustainable supplier development

Dimension	Development strategies	References
Environmental	Consulting and informing suppliers about green technology and production	Wagner and Krause (2009), Bai and Sarkis (2010)
	Setting goals for suppliers to maintain and improve the environment	Bai and Sarkis (2010), Ağan et al. (2018)
	Solving suppliers' technical environmental problems	Bai and Sarkis (2010), Narasimhan et al. (2008)
	Transferring employees with environmental expertise to suppliers	Bai and Sarkis (2010), Wagner and Krause (2009)
	Giving rewards and incentives to suppliers for their environmental performance	Bai and Sarkis (2010), Narasimhan et al. (2008), Rezaei et al. (2015)
	Helping suppliers in obtaining ISO 14000 certification	Bai and Sarkis (2010)
	Placing the environmental considerations in the long-term contracts	Bai and Sarkis (2010)
	Making commitment and providing support for green supply practices by senior management in the supplier organization	Bai and Sarkis (2010)
	Supplier involvement in designing an environmentally friendly product	Lawson et al. (2009), Bai and Sarkis, (2010)
Social	Evaluating suppliers with formal and established procedures and standards	Rezaei et al. (2015), Bowen et al. (2001), Sancha et al. (2015), Krause et al. (2000), Rezaei and Ortt (2012)
	Providing evaluation feedback to suppliers	Bowen et al. (2001), Sancha et al. (2015), Rezaei et al. (2015), Ağan et al. (2018), Rezaei and Ortt (2012)
	Performing audits for suppliers' internal management	Bowen et al. (2001), Sancha et al. (2015)
	Visiting suppliers' facilities to improve performance	Rezaei et al. (2015), Bowen et al. (2001), Sancha et al. (2015)
	Training the suppliers on the issues of sustainability	Krause et al. (2000), Narasimhan et al. (2008), Rezaei and Ortt (2012)
	Building mutual trust	Rezaei et al. (2015), Li et al. (2012)
	Training the employees of the supplier company to improve productivity	Ağan et al. (2018)
	Transferring knowledge, information, and experiences on sustainability issues	Bai and Sarkis (2010), Wagner and Krause (2009), Rezaei et al. (2015), Ağan et al. (2018)
Economic	Joint ventures	Rezaei et al. (2015), Monczka et al. (1993), Rezaei and Ortt (2012)
	Joint development and integration programs to develop new materials and products	Krause et al. (2000), Rezaei et al. (2015)

(continued)

Table 7.3 (continued)

Dimension	Development strategies	References
	Purchasing from multiple suppliers to create competition among current suppliers	Dyer and Ouchi (1993)
	Evaluating suppliers' product lines to increase productivity	Wagner and Krause (2009), Ağan et al. (2018)
	Financial evaluation and goal setting for suppliers	Ağan et al. (2018), Blome et al. (2014)

for supplier development and improvement strategies, providing feedback to suppliers after evaluation is very crucial and important, and it can help suppliers in identifying their own weaknesses.

4 Challenges in Implementing Sustainable Supplier Segmentation

Most buyer companies in the supplier relationship process only consider the stages up to supplier selection and ignore or fail in implementing the stages of segmentation, improvement, and evaluation, which are among the key stages in the supplier relationship process. Therefore, the sustainable supplier segmentation faces challenges, most of which are related to supplier development and improvement stage. The most important challenges in this respect include lack of support from senior management of both buyer and supplier companies, unwillingness to share knowledge and information, lack of sufficient knowledge related to sustainability issues, suppliers' lack of innovative capabilities, lack of effective communications between buyer and supplier, lack of open and interactive corporate culture, locational distance between buyer and supplier, economic and social differences between buyer and supplier, supplier's unwillingness to cooperate and improve relationships, and inconsistency of the buyer and supplier companies' long-term goals of corporate supply chain (Ağan et al. 2018; Rezaei 2018). By managing and improving the set of factors identified by buyer and supplier companies for development, both supplier and buyer benefit from positive performance impacts. On the other hand, some buyer companies are also unable to make a balance between sustainability dimensions for segmentation and to determine improvement strategies in each dimension and cannot provide an appropriate solution for supplier evaluation and segmentation. Companies should note that at the time of evaluating suppliers, paying simultaneous attention to all three dimensions of sustainability may affect the required allocated resources. In other words, paying attention to social and environmental indicators may lead to higher costs and affect the economic dimension of the evaluation. Therefore, the use of proper decision-making techniques can minimize the negative effects of decision-making.

5 Recommended Research Directions

Since the issue of supply chain sustainability has been debated, several researchers investigated its various aspects. Some researchers have also systematically reviewed these studies. As mentioned earlier, paying attention to the evaluation indicators and selection of sustainable suppliers has been considered in numerous studies. However, so far, no research has addressed the segmentation and development of sustainable suppliers, which is possibly due to the posed challenges. In the current era that sustainability foundations has been considered as the basis for cooperation among supply chain partners and adhering to these foundations at all stages of the supply chain has been considered as a competitive advantage, buyer-supplier cooperation across sustainability dimensions can create added value for the entire chain and finally lead to customer satisfaction. Accordingly, considering the further research, it is recommended to address the issue of sustainable supplier segmentation more technically in future studies and to direct the focus of research on issues such as sustainable supplier segmentation based on the proposed model of the present study and the impact of supplier development and improvement plans on suppliers' performance from the sustainability perspective and the overall supply chain performance.

6 Conclusion

In this chapter a model for sustainable supplier segmentation was presented which can help procurement and purchasing managers in the supply chain to evaluate and develop suppliers. In the presented segmentation model, suppliers can be divided into seven segments based on the performance indicators. In this model, the suppliers at the sustainable segment cover all three dimensions of sustainability. The sustainable segment is the most desirable part of this model, and every company tries to increase the interactions with these suppliers. Moreover, the most important supplier improvement and development strategies were presented based on sustainability dimensions, which can be used by companies to improve supplier performance and the overall supply chain performance. Using the model proposed in this chapter, the suppliers' operational weaknesses in each dimension of sustainability can be identified, and applying the improvement strategies in these dimensions can save cost, time, and manpower with maximum efficiency. Purchase, logistics, and supply chain managers in every company should provide conditions to facilitate supplier relationships and improvement after their selection, based on sustainability dimensions. By doing so, it is expected that culture of sustainability gets institutionalized across all supply chain players.

References

Adetunji, I., Price, A., Fleming, P., & Kemp, P. (2003). Sustainability and the UK construction industry—A review. In Proceedings of the Institution of Civil Engineers-Engineering Sustainability (Vol. 156, No. 4, pp. 185–199). Thomas Telford Ltd.

Ağan, Y., Acar, M. F., & Neureuther, B. (2018). The importance of supplier development for sustainability. In *Sustainable freight transport* (pp. 165–178). Cham: Springer.

Bai, C., & Sarkis, J. (2010). Green supplier development: Analytical evaluation using rough set theory. *Journal of Cleaner Production, 18*(12), 1200–1210.

Bai, C., Rezaei, J., & Sarkis, J. (2017). Multicriteria green supplier segmentation. *IEEE Transactions on Engineering Management, 64*(4), 515–528.

Björklund, M., Martinsen, U., & Abrahamsson, M. (2012). Performance measurements in the greening of supply chains. *Supply Chain Management, 17*(1), 29–39.

Blome, C., Hollos, D., & Paulraj, A. (2014). Green procurement and green supplier development: Antecedents and effects on supplier performance. *International Journal of Production Research, 52*(1), 32–49.

Bowen, F. E., Cousins, P. D., Lamming, R. C., & Farukt, A. C. (2001). The role of supply management capabilities in green supply. *Production and Operations Management, 10*(2), 174–189.

Carroll, A. B., & Buchholtz, A. K. (2002). Business and society with infotrac: Ethics and stakeholder management.

Carter, C. R., & Rogers, D. S. (2008). A framework of sustainable supply chain management: Moving toward new theory. *International Journal of Physical Distribution & Logistics Management, 38*(5), 360–387.

Chaabane, A., Ramudhin, A., & Paquet, M. (2012). Design of sustainable supply chains under the emission trading scheme. *International Journal of Production Economics, 135*(1), 37–49.

Cheung, Y. K. F., & Rowlinson, S. (2011). Supply chain sustainability: A relationship management approach. *International Journal of Managing Projects in Business, 4*(3), 480–497.

Cowell, S. J., & Parkinson, S. (2003). Localisation of UK food production: An analysis using land area and energy as indicators. *Agriculture, Ecosystems & Environment, 94*(2), 221–236.

Day, M., Magnan, G. M., & Moeller, M. M. (2010). Evaluating the bases of supplier segmentation: A review and taxonomy. *Industrial Marketing Management, 39*(4), 625–639.

Diba, S., & Xie, N. (2019). Sustainable supplier selection for Satrec Vitalait Milk Company in Senegal using the novel grey relational analysis method. *Grey Systems: Theory and Application, 9*(3), 262–294.

Dyer, J. H., & Ouchi, W. G. (1993). Japanese-style partnerships: Giving companies a competitive edge. *MIT Sloan Management Review, 35*(1), 51.

Eltayeb, T. K., Zailani, S., & Ramayah, T. (2011). Green supply chain initiatives among certified companies in Malaysia and environmental sustainability: Investigating the outcomes. *Resources, Conservation and Recycling, 55*(5), 495–506.

Faisal, M. N., Al-Esmael, B., & Sharif, K. J. (2017). Supplier selection for a sustainable supply chain: Triple bottom line (3BL) and analytic network process approach. *Benchmarking: An International Journal, 24*(7), 1956–1976.

Fernando, Y., & Saththasivam, G. (2017). Green supply chain agility in EMS ISO 14001 manufacturing firms: Empirical justification of social and environmental performance as an organisational outcome. *International Journal of Procurement Management, 10*(1), 51–69.

Garg, C. P., & Sharma, A. (2018). Sustainable outsourcing partner selection and evaluation using an integrated BWM–VIKOR framework. *Environment, Development and Sustainability*, 1–29.

Glock, C. H., Grosse, E. H., & Ries, J. M. (2017). Reprint of "Decision support models for supplier development: Systematic literature review and research agenda". *International Journal of Production Economics, 194*, 246–260.

Govindan, K., Khodaverdi, R., & Jafarian, A. (2013). A fuzzy multi criteria approach for measuring sustainability performance of a supplier based on triple bottom line approach. *Journal of Cleaner Production, 47*, 345–354.

Hadjikhani, A., & LaPlaca, P. (2013). Development of B2B marketing theory. *Industrial Marketing Management, 42*(3), 294–305.

Handfield, R. B., Walton, S. V., Seegers, L. K., & Melnyk, S. A. (1997). 'Green' value chain practices in the furniture industry. *Journal of Operations Management, 15*(4), 293e315.

Holt, D., & Ghobadian, A. (2009). An empirical study of green supply chain management practices amongst UK manufacturers. *Journal of Manufacturing Technology Management, 20*(7), 933–956.

Jin, Y., Ryan, J. K., & Yund, W. (2013). Two stage procurement processes with competitive suppliers and uncertain supplier quality. *IEEE Transactions on Engineering Management, 61* (1), 147–158.

Kannan, V. R., & Tan, K. C. (2002). Supplier selection and assessment: Their impact on business performance. *Journal of Supply Chain Management, 38*(3), 11–21.

Kraljic, P. (1983). Purchasing must become supply management. *Harvard Business Review, 61*(5), 109–117.

Krause, D. R., Scannell, T. V., & Calantone, R. J. (2000). A structural analysis of the effectiveness of buying firms' strategies to improve supplier performance. *Decision Sciences, 31*(1), 33–55.

Lawson, B., Cousins, P. D., Handfield, R. B., & Petersen, K. J. (2009). Strategic purchasing, supply management practices and buyer performance improvement: An empirical study of UK manufacturing organizations. *International Journal of Production Research, 47*(10), 2649–2667.

Li, W., Humphreys, P. K., Yeung, A. C., & Cheng, T. C. E. (2012). The impact of supplier development on buyer competitive advantage: A path analytic model. *International Journal of Production Economics, 135*(1), 353–366.

Luthra, S., Govindan, K., Kannan, D., Mangla, S. K., & Garg, C. P. (2017). An integrated framework for sustainable supplier selection and evaluation in supply chains. *Journal of Cleaner Production, 140*, 1686–1698.

Marshall, D., McCarthy, L., McGrath, P., & Claudy, M. (2015). Going above and beyond: How sustainability culture and entrepreneurial orientation drive social sustainability supply chain practice adoption. *Supply Chain Management: An International Journal, 20*(4), 434–454.

Martínez-Blanco, J., Lehmann, A., Muñoz, P., Antón, A., Traverso, M., Rieradevall, J., & Finkbeiner, M. (2014). Application challenges for the social Life Cycle Assessment of fertilizers within life cycle sustainability assessment. *Journal of Cleaner Production, 69*, 34–48.

Martins, C. L., & Pato, M. V. (2019). Supply chain sustainability: A tertiary literature review. *Journal of Cleaner Production, 225*, 995–1016.

Memon, M. S., Lee, Y. H., & Mari, S. I. (2015). Group multi-criteria supplier selection using combined grey systems theory and uncertainty theory. *Expert Systems with Applications, 42* (21), 7951–7959.

Monczka, R. M., Trent, R. J., & Callahan, T. J. (1993). Supply base strategies to maximize supplier performance. *International Journal of Physical Distribution & Logistics Management, 23*(4), 42–54.

Narasimhan, R., Mahapatra, S., & Arlbjørn, J. S. (2008). Impact of relational norms, supplier development and trust on supplier performance. *Operations Management Research, 1*(1), 24–30.

Olsen, R. F., & Ellram, L. M. (1997). A portfolio approach to supplier relationships. *Industrial Marketing Management, 26*(2), 101–113.

Ortas, E., Moneva, M., & Álvarez, I. (2014). Sustainable supply chain and company performance: A global examination. *Supply Chain Management: An International Journal, 19*(3), 332–350.

Parasuraman, A. (1980). Vendor segmentation: An additional level of market segmentation. *Industrial Marketing Management, 9*(1), 59–62.

Rezaei, J. (2018). Sustainable supplier selection: A process view. *Handbook on the Sustainable Supply Chain*, 136–147.

Rezaei, J., & Ortt, R. (2012). A multi-variable approach to supplier segmentation. *International Journal of Production Research, 50*(16), 4593–4611.

Rezaei, J., & Ortt, R. (2013). Multi-criteria supplier segmentation using a fuzzy preference relations based AHP. *European Journal of Operational Research, 225*(1), 75–84.

Rezaei, J., Fahim, P. B., & Tavasszy, L. (2014). Supplier selection in the airline retail industry using a funnel methodology: Conjunctive screening method and fuzzy AHP. *Expert Systems with Applications, 41*(18), 8165–8179.

Rezaei, J., Wang, J., & Tavasszy, L. (2015). Linking supplier development to supplier segmentation using Best Worst Method. *Expert Systems with Applications, 42*(23), 9152–9164.

Sancha, C., Gimenez, C., Sierra, V., & Kazeminia, A. (2015). Does implementing social supplier development practices pay off? *Supply Chain Management: An International Journal, 20*(4), 389–403.

Segura, M., & Maroto, C. (2017). A multiple criteria supplier segmentation using outranking and value function methods. *Expert Systems with Applications, 69*, 87–100.

Seuring, S., & Müller, M. (2008). From a literature review to a conceptual framework for sustainable supply chain management. *Journal of Cleaner Production, 16*(15), 1699–1710.

Su, C. M., Horng, D. J., Tseng, M. L., Chiu, A. S., Wu, K. J., & Chen, H. P. (2016). Improving sustainable supply chain management using a novel hierarchical grey-DEMATEL approach. *Journal of Cleaner Production, 134*, 469–481.

Svensson, G. (2004). Supplier segmentation in the automotive industry: A dyadic approach of a managerial model. *International Journal of Physical Distribution & Logistics Management, 34* (1), 12–38.

Talluri, S., Narasimhan, R., & Chung, W. (2010). Manufacturer cooperation in supplier development under risk. *European Journal of Operational Research, 207*(1), 165–173.

Tavana, M., Yazdani, M., & Di Caprio, D. (2017). An application of an integrated ANP–QFD framework for sustainable supplier selection. *International Journal of Logistics Research and Applications, 20*(3), 254–275.

Vachon, S., & Klassen, R. D. (2008). Environmental management and manufacturing performance: The role of collaboration in the supply chain. *International Journal of Production Economics, 111*(2), 299–315.

Vachon, S., & Mao, Z. (2008). Linking supply chain strength to sustainable development: A country-level analysis. *Journal of Cleaner Production, 16*(15), 1552–1560.

Wagner, S. M., & Krause, D. R. (2009). Supplier development: Communication approaches, activities and goals. *International Journal of Production Research, 47*(12), 3161–3177.

Wang, H. S., Che, Z. H., & Wang, M. J. (2009). A three-phase integrated model for product configuration change problems. *Expert Systems with Applications, 36*(3), 5491–5509.

Wang, Z., Subramanian, N., Gunasekaran, A., Abdulrahman, M. D., & Liu, C. (2015). Composite sustainable manufacturing practice and performance framework: Chinese auto-parts suppliers′ perspective. *International Journal of Production Economics, 170*, 219–233.

Wu, M. W., & Shen, C. H. (2013). Corporate social responsibility in the banking industry: Motives and financial performance. *Journal of Banking & Finance, 37*(9), 3529–3547.

Xu, Z., Qin, J., Liu, J., & Martínez, L. (2019). Sustainable supplier selection based on AHPSort II in interval type-2 fuzzy environment. *Information Sciences, 483*, 273–293.

Yusuf, Y. Y., Gunasekaran, A., Musa, A., El-Berishy, N. M., Abubakar, T., & Ambursa, H. M. (2013). The UK oil and gas supply chains: An empirical analysis of adoption of sustainable measures and performance outcomes. *International Journal of Production Economics, 146*(2), 501–514.

Zhu, Q., Sarkis, J., & Lai, K. H. (2007). Green supply chain management: Pressures, practices and performance within the Chinese automobile industry. *Journal of Cleaner Production, 15* (11–12), 1041–1052.

Zhu, Q., Sarkis, J., & Lai, K. H. (2008). Confirmation of a measurement model for green supply chain management practices implementation. *International Journal of Production Economics, 111*(2), 261–273.

Zimmer, K., Fröhling, M., & Schultmann, F. (2016). Sustainable supplier management–a review of models supporting sustainable supplier selection, monitoring and development. *International Journal of Production Research, 54*(5), 1412–1442.

Chapter 8
Evaluation of Manufacturing Organizations Ability to Overcome Internal Barriers to Green Innovations

Himanshu Gupta and Mukesh Kumar Barua

Abstract Manufacturing organizations are under constant pressure from regulatory bodies to reduce the environmental impact of their products and processes. The stiff competition is also among manufacturing organizations to improve their social image. Manufacturing organizations, especially in developing countries, are lagging behind their counterparts in developed countries in terms of commitment toward the environment. Green innovations at their organizations are the economic and viable solution to overcome this challenge. However, the manufacturing organizations are struggling to green their operations due to numerous barriers like lack of resources and unavailability of financial and human capital. It is often the internal barriers on which organizations have some control. This study identifies internal barriers to green innovation through literature review. A hybrid of three different methodologies is employed. First, ISM is employed to identify the relationship between these barriers. ISM also helps in identifying driving barriers that have the most effect on the system. Next, using BWM, the driving barriers are ranked. In the third step, VIKOR methodology is applied to rank the ability of manufacturing organizations in overcoming these barriers.

Keywords Green innovation · ISM · Best-Worst Method · VIKOR · Internal barriers

---

H. Gupta (✉)
Department of Management Studies, Indian Institute of Technology (Indian School of Mines) Dhanbad, Dhanbad, India

M. K. Barua
Department of Management Studies, Indian Institute of Technology Roorkee, Roorkee, India

J. Rezaei (ed.), *Strategic Decision Making for Sustainable Management of Industrial Networks*, Greening of Industry Networks Studies 8,
https://doi.org/10.1007/978-3-030-55385-2_8

1 Introduction

Rapid industrial growth and improper resource management in the past decade has led to imbalance in demand and supply of resources; also industrial growth has caused environmental degradation which is a major concern worldwide (Tam et al. 2006; Zhe et al. 2016). Few years back investment in environmental activities and greening of operations were considered unnecessary burden by the companies, but recently due to stringent government regulations and competition to attract new customers on the basis of organizations, green image has led to changing in thinking and pattern of investment for organizations (Porter and Van der Linde 1995). Due to resource constraints and demand to provide different products at low cost, embracing green innovation is necessity for organizations, and green innovation is an important tool for sustainable development of organization (Chang 2011). Green innovation is conceptualized as those activities which are green and helps the organizations to optimize internal resources thus enhancing their ability to produce green products and services. These activities must comply with environmental regulations and market demand and use green and sustainable technologies by involving all the stakeholders (Gupta and Barua 2017, 2018; Fernando et al. 2019). Green innovation is an important pillar for major sustainable shift across various manufacturing organizations as well as for green supply chains (Dubey et al. 2015; Zhu and Sarkis 2007). Green innovation requires human resources which are capable to learn, organize, execute, and monitor the eco-efficient activities at the organizations (Song and Wang 2018). Organizations' orientation toward green innovation can help them to solve environmental degradation-related issues, and addressing those issues will help the organizations to provide better service and products to the customers (Eiadat et al. 2008). The literature also supports the fact that the green innovation is instrumental for improving performance of the organizations.

The choice of an organization to implement the type of green innovation and technologies lies with its organizational goals and operational conditions. The success of green innovation technologies depends upon a variety of factors like external factors that include collaboration with supply chain partners and internal factors like organizations capacity to adopt green technology, technical competencies, etc. (Hermann et al. 2016). External factors are difficult to control directly by the organizations as they depend on a number of external factors; on the other hand internal barriers are equally complex due to their micro level interaction and dependency on each other (Song et al. 2018). The internal barriers are still easy to control and manage and hence should be given priority. Adopting green innovation practices is not easy task and organizations need to start building its internal capabilities and overcome internal barriers to green innovation capabilities first rather than depending on external help for greening its operations. In this context, this study has the following objectives:

- To identify the internal barriers to green innovation
- To identify the hierarchal structure and relationship among these barriers

- To prioritize these internal barriers to green innovation
- To evaluate organizations on the basis of these barriers

This chapter is divided into the following sections: the next section gives a brief background of the topic and deals with identification of barriers through literature, the third section deals with the details of three-phase methodology employed in this study, fourth section deals with case analysis and presents results, and the last section discusses results and presents conclusions and future recommendations.

2 Green Innovation and Sustainability

Green and sustainable developments are the major goals of organizations across the world in order to meet the sustainability goals of 2030. From the context of environment, the term "green" is mainly related to the products or technologies which generate no or minimum waste and pollution, and the term "sustainable" refers to the lesser and lesser use of resources and also production or processing at the minimum possible cost (Liao and Wang 2018). Although the term green mainly relates to environmental aspects, green innovation caters to all the three pillars of sustainability, viz., environment, economic, and social. Green innovation which involves both "green process" and "green product" innovations involves handling various issues like energy saving, material saving, reduction in environmental pollution, less costly and innovative production processes, packaging design, waste management, and recycling. By tackling these issues, green innovation can help organizations achieve their sustainability related goals (Chen et al. 2006; Saunila et al. 2018). Green innovation can be defined as actions related to pollution prevention, recycling of waste, green design of products, technologies for energy saving, and environmental management practices aimed at waste and cost reduction and acquiring new profits (Chen et al. 2006). Recently many studies have been carried out in the field of green innovation. Chen et al. (2006) studied the impact of green innovation performance on organizations competitive environment in electronics industry of Taiwan. Both green product innovation and green process innovation performance are found to be positively related with corporate competitive advantage. However, as suggested by various authors, there are numerous barriers to green innovation which are hampering this competitive advantage of greening the processes (Gupta and Barua 2018; Xia et al. 2019). Majority of these barriers are internal to the organizations and can be controlled more easily than the external barriers. After thorough review of literature, 11 internal barriers to green innovation are identified and are presented in Table 8.1.

Table 8.1 Organizations internal barriers to green innovation

Listed barriers	Description	Supporting literature
Lack of top management commitment (BI01)	It refers to lack of willingness of top management and owners of the organization to incorporate green technologies	Mangla et al. (2017) and Gupta and Barua (2018)
Resistance from employees (BI02)	It refers to resistance by employees to the adoption of new technologies due to the fear that it will upset their status quo	Lin and Ho (2008), De Jesus and Mendonça (2018), and Gupta and Barua (2018)
Lack of skilled human resources (BI03)	It refers to lack of manpower and staff that have necessary skills to adopt green innovation	Collins et al. (2007), Lin and Ho (2008), and Xia et al. (2019)
Insufficient motivation and rewards (BI04)	It refers to lack of reward schemes for employees for creative and innovative thinking and bringing change related to greening of the products and processes in the organization	Madrid-Guijarro et al. (2009)
Lack of perceived benefits (BI05)	It refers to lack of belief among employees and managers about the probable benefits of green products	Walker et al. (2008), Govindan et al. (2014), De Jesus and Mendonça (2018), Gupta and Barua (2018), and Xia et al. (2019)
Lag between design and diffusion (BI06)	It refers to inability of the organization to incorporate the design modifications done into their products and processes	De Jesus and Mendonça (2018)
Insufficient knowledge (BI07)	It refers to lack of knowledge regarding green technologies and processes among the employees and managers of the organization	Madrid-Guijarro et al. (2009), Mangla et al. (2017), De Jesus and Mendonça (2018), and Xia et al. (2019)
Lack of required infrastructure (BI08)	It refers to lack of infrastructure to process and produce green products. The production machinery might be able to process steel but might not be able to process plastic properly	Dubey et al. (2017) and Xia et al. (2019)
Lack of technologies (BI09)	It refers to lack of technologies within the organization to carry out green innovation related activities and also for recycling and reuse of products	Del Río et al. (2010), Gupta and Barua (2018), and Xia et al. (2019)
Lack of financial support (BI10)	It refers to insufficient funds for carrying out activities related to greening of the products and processes	Cecere et al. (2018) and De Jesus and Mendonça (2018)
Insufficient R&D facilities (BI11)	It refers to lack of research and development facilities at the organization for carrying out activities related to the greening of the operations	Perron (2005) and Pawanchik and Sulaiman (2010)

2.1 Research Gaps and Problem Formulation

Most of the past studies on green innovation have focused on few factors at a time and are often limited to finding the effect of one barrier on other; there is no such study to first find the interdependence relationship among barriers and also to rank these barriers. Further most of the studies have taken both internal and external barriers that limit the adoption of green innovations in the organization, and there is no such study on firms' internal barriers to green innovation implementation. Also there is no study to evaluate organizations performance on the basis of internal barriers to green innovation. This study aims to first find the hierarchal structure and dependence relationship among identified barriers and then rank these barriers and finally using weights of these barriers evaluate organizations and rank them. Organizations and decision-makers need to have a framework through which they can identify barriers to green innovation in general and also specific to their organizations and also to evaluate their performance on these barriers. This chapter presents that framework for decision-makers to evaluate their organizations on green and sustainability innovation barriers.

3 Methodology

This study involves a three-phase multi-criteria decision-making methodology (MCDM). MCDM techniques deal with the issue of obtaining the best possible solution from a set of multiple criteria. Often the problem with large number of criteria is that no single criteria is available that can dominate all the criteria; thus decision-makers don't have a clear-cut solution to their problem, and hence they require a structured methodology to find the best solution (Hafezalkotob et al. 2019). The major objective if this research is to rank the manufacturing organizations on the basis of their performance to handle internal barriers to green innovation. For selection of internal barriers for this framework, authors want to select the barriers with highest driving power. To meet this objective, this study has selected three different methodologies, ISM (interpretive structural modeling) is selected for obtaining a hierarchical structure of the finalized barriers and also categorizing these barriers according to their driving power. Next step uses BWM (Best-Worst Method) for ranking the selected barriers (after first phase) based upon their weights. BWM is used because it has emerged as the most popular MCDM technique in the recent few years, and it has advantage over other popular techniques like AHP (Analytical Hierarchical Processing) that it requires lesser number of pairwise comparisons for getting the desired results. Third phase involves application of VIKOR methodology for ranking of alternatives, VIKOR is used due to its ability to provide a compromise solution in case of a conflicting situation where a number of alternatives are evaluated. The methodology is briefly represented in Fig. 8.1.

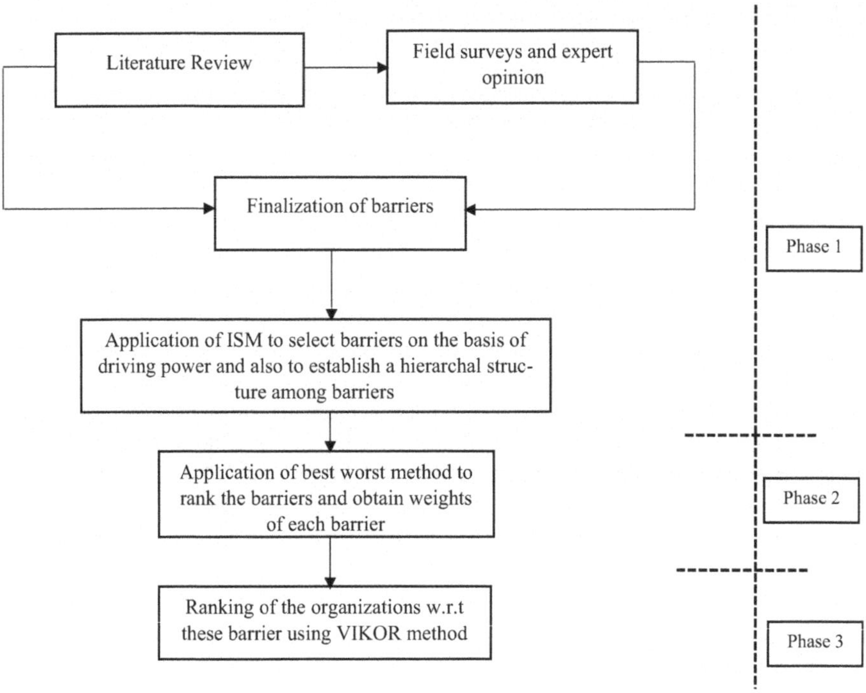

Fig. 8.1 Solution methodology

3.1 ISM (Interpretive Structural Modeling)

ISM is a powerful technique for determining the interrelationship among variables and is backed by the concepts of reachability matrix and transitivity relationship among variables (Warfield 1974). These two concepts can be understood using the following example: suppose we have three variables "i," "j," and "k"; here we assume that variable "i" is driving variable "j" and variable "k" is driven by "j" as well as "i"; then the relationship between "i" and "j" as well as "j" and "k" can be shown by two arrows with arrowhead pointing toward "j" and "k," respectively. In this case, the relationship between "i" and "j" need not be shown using an arrow. Reachability grid helps in making the different levels in ISM. The different steps followed in ISM are discussed below (Al-Muftah et al. 2018; Hughes et al. 2019; Malek and Desai 2019).

Step 1: Development of a Structural Self-Interaction Matrix (SSIM)

In order to analyze the relationship among factors considered in the study, a contextual relationship of "leads to" type is chosen. For identification of contextual relationship, a group of experts are consulted. Four different letters are used to represent the direction of the relationship between the variables *i* and *j*:

V: It indicates that variable *i* directly leads variable *j*.
A: It indicates that variable *j* directly leads variable *i*.
X: It indicates that both the variables *i* and *j* influence/lead each other.
O: It indicates that variables *i* and *j* are not leading each other and also have no relationship.

Step 2: Construction of Reachability Matrix

The next step is formation of an initial reachability matrix from SSIM. It is done by replacing the values of "V, A, X, and O" with either 0 or 1 as per the rules mentioned below:

- In case the respondent has marked V for (*i*, *j*) cell in the SSIM, then the reachability matrix will be prepared with (*i*, *j*) cell as 1 and the (*j*, *i*) cell as 0.
- In case the respondent has marked A for (*i*, *j*) cell in the SSIM, reachability matrix will be prepared by entering the (*i*, *j*) cell as 0 and the (*j*, *i*) cell as 1.
- In case the respondent has marked X for (*i*, *j*) cell in the SSIM, then the reachability matrix will be prepared by entering 1in (*i*, *j*) as well as (*j*, *i*) cell.
- In case the respondent has marked O for the (*i*, *j*) cell in the SSIM, then the reachability matrix is prepared by entering 0 both in (*i*, *j*) cell as well as (*j*, *i*) cell.

After all the above rules are applied, final reachability matrix is obtained by applying transitivity rules; the transitivity rules say that if some factor "A" is related or influencing factor "B" and factor "B" is influencing factor "C," then factor "A" must also influence factor "C." The difference between initial and final reachability matrices can be determined by * symbol on cells for which transitivity rule is applied.

Step 3: Partitioning of Factors into Different Levels

The next step is to obtain reachability and antecedent set for all the variables in the study. The reachability set for an individual factor includes itself and other factors which it could assist to accomplish, while the antecedent set consists of the factors themselves and other factors which may help in achieving it. The reachability set is prepared by taking the driving power values from final reachability matrix; all the variables having driving power value as "1" are included in reachability set. Similarly, antecedent set is prepared by considering the dependence power for that particular variable; all the corresponding variables having dependence power value as "1" are included in antecedent set. The intersection set represents the common values of reachability and antecedent sets. Finally, the levels are decided for a variable if for that particular variable, both reachability set and intersection set are common. The top level of ISM hierarchy is assigned to the variables having same reachability and intersection sets. The same approach is followed in achieving other levels of ISM hierarchy by eliminating the variable(s) that have already been assigned any level and repeating the above process.

Step 4: Formation of ISM-Based Model

Final reachability matrix and level partition table is used to generate the final hierarchical structure. All the variables which are at level 1 are placed at the bottom of the hierarchy and are considered most important and also drives the whole system. The relationship between factor *i* and *j* is denoted by an arrow where direction of the arrow shows that two factors are related. The resulting graph obtained by representing the relationship by arrows and level partitioning is known as "directed graph" or "digraph," which is also the final hierarchical structure.

3.2 Best-Worst Method

BWM is a very strong MCDM technique and is widely used by researchers all over the world for decision-making and problem-solving. It is developed by Rezaei (2015, 2016) and widely used by like Gupta and Barua 2016 (technological innovation enablers ranking); Rezaei et al. 2016 (green supplier selection); Gupta and Barua 2017 (green supplier selection); Gupta 2018 (airport evaluation based on service quality); Ahmadi et al. 2017 (social sustainability in supply chains); Zhao et al. 2018 (benefit evaluation of eco-industrial parks from circular economy and sustainability perspective); Malek and Desai (sustainable manufacturing barriers analysis). The steps as given by Rezaei (2015, 2016) are explained below:

Step 1: Selection of criteria (barriers) for analysis.
Through literature review internal barriers to green innovation are finalized.
Step 2: Among finalized criteria best and the worst criteria is chosen by experts.
Step 3: Next each reference ratings on the scale of 1–9 are sought from experts for best to others and others to worst criteria.
Step 4: Optimized weights ($w_1^*, w_2^*, \ldots\ldots, w_n^*$) for all the criteria is calculated next.

The objective is to obtain the weights of criteria so that the maximum absolute differences for all j can be minimized for

$$\left\{\left|w_B - a_{Bj}w_j\right|, \left|w_j - a_{jW}w_W\right|\right\}.$$

This minimax model will be obtained:

$$\begin{gathered}\min\ \max\ \left\{\left|w_B - a_{Bj}w_j\right|, \left|w_j - a_{jW}w_W\right|\right\} \\ \text{s.t.} \sum_j w_j = 1 \\ w_j \geq 0, \text{for all } j\end{gathered} \tag{8.1}$$

Model (8.1) is transformed into a linear model; the model is shown below:

$$\min\ \xi^L$$

s.t.

$$\left|w_B - a_{Bj}w_j\right| \leq \xi^L, \text{for all } j$$

$$\left|w_j - a_{jW}w_W\right| \leq \xi^L, \text{for all } j$$

$$\sum_j w_j = 1$$

$$w_j \geq 0, \text{for all } j \tag{8.2}$$

Model (8.2) can be solved to obtain optimal weights ($w_1^*, w_2^*, \ldots\ldots, w_n^*$) and optimal value$\xi^L$.

Consistency (ξ^L) of attribute comparisons close to 0 is desired (Rezaei 2016).

3.3 Evaluating Organizations Using VIKOR

VIKOR method as introduced by Opricovic (1998) has the advantage to provide optimized solutions in case of complex and conflicting situations and also in cases where criteria have different units of measurement; it provides an optimized solution that is closest to an ideal solution using compromise priority approach. The steps of VIKOR methodology are presented below:

Step 1: Obtain a pairwise matrix of criteria and alternatives using scale mentioned in Table 8.2.

Step 2: After that using Eq. (8.3) the average decision matrix is obtained

Table 8.2 Linguistic scale for pairwise comparison for VIKOR methodology

Scale for VIKOR methodology	
Linguistic variables	Importance rating
Least important	1
Moderately important	2
Strongly important	3
Very strongly important	4
Extremely important	5

$$F = \frac{1}{k} \sum_{k=1}^{k} F_k \tag{8.3}$$

Step 3: Using Eqs. (8.4) and (8.5), best f_b^* and the worst f_b^- values of all the criteria, $b = 1, 2, \ldots .n$ are obtained

$$f_b^* = \text{Max}\ (f_{ab}) \tag{8.4}$$

$$f_b^- = \text{Min}\ (f_{ab}) \tag{8.5}$$

where f_b^* is the positive ideal solution and f_b^- is the negative ideal solution for the bth attribute.

Step 4: Compute the S_a and R_a values for a = 1, 2, … .m using Eqs. (8.6) and (8.7).

$$S_a = \sum_{b=1}^{n} W_b \left[\left(f_b^* - f_{ab} \right) / \left(f_b^* - f_b^- \right) \right] \tag{8.6}$$

$$R_a = \text{Max}_b \left[W_b \left(f_b^* - f_{ab} \right) / \left(f_b^* - f_b^- \right) \right] \tag{8.7}$$

where S_a and R_a are the distance of ath alternative from positive ideal solution and negative ideal solution, respectively, and W_b represents the weights of the criteria.

Step 5: Using Eq. (8.8) compute the scores forQ_a.

$$Q_a = v\left(\frac{S_a - S^*}{S^- - S^*}\right) + (1 - v)\left(\frac{R_a - R^*}{R^- - R^*}\right) \tag{8.8}$$

where $S^- = \text{Max}_a S_a$, $S^* = \text{Min}_a S_a$, $R^- = \text{Max}_a R_a$, $R^* = \text{Min}_a R_a$ and v denotes the weightage of maximum set utility and is taken as 0.5 in this study.

Step 6: Using Q_a values alternatives are ranked.

Step 7: Alternatives are ranked based on minimum Q_a values obtained subject to simultaneously satisfying two conditions:

Condition 1: Q(A(1)) is chosen if Q(A(2)) – Q(A(1)) ≥ 1/n-1 where A(2) is the alternative that has got the second rank in the analysis and n is the total alternatives.

Condition 2: Q(A(1)) also obtains the first rank according to both S_a and R_a values.

Step 8: Alternative that obtained a minimum score in Q_a is ranked first.

4 Case Analysis and Results

In the current study, in order to evaluate the performance of manufacturing organizations, four different organizations are taken, and one expert each from the participating organizations is asked to participate in the survey. The methodology used for collecting data from the experts is panel consensus method for ISM and BWM. For VIKOR, individual ratings are obtained from each expert, and results are aggregated by taking average of their ratings. The panel consensus method involves inviting all the experts together for a panel discussion, where they can discuss and deliberate upon common option (rating) after mutual consensus of each of the expert. It has the advantage that it is time-saving in the sense that all the experts are at one place and also decision is with the consent of each of the expert. Also, it is convenient for researcher to collect data from a single place rather than moving to different places to meet different experts. The aggregate method for data collection is adopted for VIKOR because in the organizations are different, and it is very difficult to get same opinion for all the experts for four different organizations performance, and hence panel consensus will not work here. This method has advantage over panel consensus that it allows the decision-maker to make decisions individually and not under the influence of other experts who might be senior to him and might cause his influence on the decision-maker thus creating a bias in the results. The profiles of the various experts are as follows: Expert 1 is a senior manager of operations and maintenance at a leading automobile company and has around 8 years of experience; Expert 2 is an associate general manager of supply chain and procurement at a leading automobile company with about 14 years of experience; Expert 3 is a manager of operations and also manager in charge of environmental activities at a leading plastic component manufacturing organization, he has about 10 years of experience; and Expert 4 is a senior manager of procurement at a leading steel manufacturing organization with 12 years of experience.

Through extensive literature review, 11 internal barriers to green innovation are identified, and as mentioned in Section 3.1, the panel of experts is asked to give their preference ratings to form the structural self-interaction matrix as shown in Table 8.3.

Using the rules mentioned in Sect. 3.1, the structural self-interaction matrix is converted into initial reachability matrix as shown in Table 8.4.

After obtaining initial reachability matrix, the rules of transitivity are applied to obtain a final reachability matrix as shown in Table 8.5. This final reachability matrix is also used to obtain driving and dependence power of the barriers by summating the number of "1" in rows and columns, respectively.

The next step is to do level portioning. Level partitioning is done by obtaining reachability and antecedent sets as explained in Section 3. The reachability set is prepared by taking the driving power values from final reachability matrix; all the variables having driving power value as "1" are included in reachability set. Similarly, antecedent set is prepared by considering the dependence power for that particular variable; all the corresponding variables having dependence power value

Table 8.3 Structural self-interaction matrix

	BI11	BI10	BI09	BI08	BI07	BI06	BI05	BI04	BI03	BI02	BI01
BI01	X	A	A	X	X	O	V	V	V	X	
BI02	A	O	A	V	V	O	V	X	O		
BI03	V	V	O	V	V	V	V	X			
BI04	A	A	A	V	V	V	V				
BI05	A	A	A	A	A	V					
BI06	A	A	A	A	A						
BI07	X	A	A	X							
BI08	A	A	O								
BI09	O	X									
BI10	V										
BI11											

Table 8.4 Initial Reachability Matrix

	BI01	BI02	BI03	BI04	BI05	BI06	BI07	BI08	BI09	BI10	BI11
BI01	1	1	1	1	1	0	1	1	0	0	1
BI02	1	1	0	1	1	0	1	1	0	0	0
BI03	0	0	1	1	1	1	1	1	0	1	1
BI04	0	1	1	1	1	1	1	1	0	0	0
BI05	0	0	0	0	1	1	0	0	0	0	0
BI06	0	0	0	0	0	1	0	0	0	0	0
BI07	1	0	0	0	1	1	1	1	0	0	1
BI08	1	0	0	0	1	1	1	1	0	0	0
BI09	1	1	0	1	1	1	1	0	1	1	0
BI10	1	0	0	1	1	1	1	1	1	1	1
BI11	1	1	0	1	1	1	1	1	0	0	1

as "1" are included in antecedent set. The intersection set represents the common values of reachability and antecedent sets. Finally, the levels are decided for a variable if for that particular variable, both reachability set and intersection set is common. For better understanding of level partitioning, Table 8.6 represents the level partitioning of various barriers (Tables 8.7, 8.8, and 8.9).

This level partitioning is used to construct the hierarchical structure as represented in Fig. 8.3.

The values obtained through driving and dependence power of each of the barrier are used to plot a graph for MICMAC analysis, where driving power is plotted on X-axis and dependence power on Y-axis as shown in Fig. 8.2.

All the barriers above the half of Y-axis (i.e., top two quadrants) are considered to have high driving power on the whole system. Especially the barriers in the top left quadrant have very high driving power and low dependence power and are considered most important for the system. In this case lack of technologies (BI09) is the only barrier in top left quadrant and hence has the most impact on the whole system (Fig. 8.3).

Table 8.5 Final reachability matrix

	BI01	BI02	BI03	BI04	BI05	BI06	BI07	BI08	BI09	BI10	BI11	Driving Power
BI01	1	1	1	1	1	1*	1	1	0	1*	1	10
BI02	1	1	1*	1	1	1*	1	1	0	0	1*	9
BI03	1*	1*	1	1	1	1	1	1	1*	1	1	11
BI04	1*	1	1	1	1	1	1	1	0	1*	1*	10
BI05	0	0	0	0	1	1	0	0	0	0	0	2
BI06	0	0	0	0	0	1	0	0	0	0	0	1
BI07	1	1*	1*	1*	1	1	1	1	0	0	1	9
BI08	1	1*	1*	1*	1	1	1	1	0	0	1*	9
BI09	1	1	1*	1	1	1	1	1*	1	1	1*	11
BI10	1	1*	1*	1	1	1	1	1	1	1	1	11
BI11	1	1	1*	1	1	1	1	1	0	1*	1	10
Dependence Power	9	9	9	9	10	11	9	9	3	6	9	

Table 8.6 Iteration 1 for level partitioning

Variable	Reachability set	Antecedent set	Intersection set	Level
BI01	1, 2, 3, 4, 5, 6, 7, 8, 10, 11	1, 2, 3, 4, 7, 8, 9, 10, 11	1, 2, 3, 4, 7, 8, 10, 11	
BI02	1, 2, 3, 4, 5, 6, 7, 8, 11	1, 2, 3, 4, 7, 8, 9, 10, 11	1, 2, 3, 4, 7, 8, 11	
BI03	1, 2, 3, 4, 5, 6, 7, 8, 9, 10, 11	1, 2, 3, 4, 7, 8, 9, 10, 11	1, 2, 3, 4, 7, 8, 9, 10, 11	
BI04	1, 2, 3, 4, 5, 6, 7, 8, 10, 11	1, 2, 3, 4, 7, 8, 9, 10, 11	1, 2, 3, 4, 7, 8, 10, 11	
BI05	5, 6	1, 2, 3, 4, 5, 7, 8, 9, 10, 11	5	
BI06	6	1, 2, 3, 4, 5, 6, 7, 8, 9, 10, 11	6	I
BI07	1, 2, 3, 4, 5, 6, 7, 8, 11	1, 2, 3, 4, 7, 8, 9, 10, 11	1, 2, 3, 4, 7, 8, 11	
BI08	1, 2, 3, 4, 5, 6, 7, 8, 11	1, 2, 3, 4, 7, 8, 9, 10, 11	1, 2, 3, 4, 7, 8, 11	
BI09	1, 2, 3, 4, 5, 6, 7, 8, 9, 10, 11	3, 9, 10	3, 9, 10	
BI10	1, 2, 3, 4, 5, 6, 7, 8, 9, 10, 11	1, 3, 4, 9, 10, 11	1, 3, 4, 9, 10, 11	
BI11	1, 2, 3, 4, 5, 6, 7, 8, 10, 11	1, 2, 3, 4, 7, 8, 9, 10, 11	1, 2, 3, 4, 7, 8, 10, 11	

Table 8.7 Iteration 2 for level portioning

Variable	Reachability set	Antecedent set	Intersection set	Level
BI01	1, 2, 3, 4, 5, 7, 8, 10, 11	1, 2, 3, 4, 7, 8, 9, 10, 11	1, 2, 3, 4, 7, 8, 10, 11	
BI02	1, 2, 3, 4, 5, 7, 8, 11	1, 2, 3, 4, 7, 8, 9, 10, 11	1, 2, 3, 4, 7, 8, 11	
BI03	1, 2, 3, 4, 5, 7, 8, 9, 10, 11	1, 2, 3, 4, 7, 8, 9, 10, 11	1, 2, 3, 4, 7, 8, 9, 10, 11	
BI04	1, 2, 3, 4, 5, 7, 8, 10, 11	1, 2, 3, 4, 7, 8, 9, 10, 11	1, 2, 3, 4, 7, 8, 10, 11	
BI05	5	1, 2, 3, 4, 5, 7, 8, 9, 10, 11	5	II
BI07	1, 2, 3, 4, 5, 7, 8, 11	1, 2, 3, 4, 7, 8, 9, 10, 11	1, 2, 3, 4, 7, 8, 11	
BI08	1, 2, 3, 4, 5, 7, 8, 11	1, 2, 3, 4, 7, 8, 9, 10, 11	1, 2, 3, 4, 7, 8, 11	
BI09	1, 2, 3, 4, 5, 7, 8, 9, 10, 11	3, 9, 10	3, 9, 10	
BI10	1, 2, 3, 4, 5, 7, 8, 9, 10, 11	1, 3, 4, 9, 10, 11	1, 3, 4, 9, 10, 11	
BI11	1, 2, 3, 4, 5, 7, 8, 10, 11	1, 2, 3, 4, 7, 8, 9, 10, 11	1, 2, 3, 4, 7, 8, 10, 11	

The next step in ISM analysis is to form a hierarchal structure of the barriers by partitioning the barriers in different levels. Using the rules mentioned in Sect. 3.1, the barriers are portioned into different levels. The barriers that come at level I are place at the bottom of the hierarchy and are considered to have most impact on all other barriers, i.e., they are considered to drive the whole system. In this case lack of technologies (BI09) and lack of financial support (BI10) are at level I and hence are considered to drive the whole system.

In the second phase of the analysis, the objective is to rank the internal barriers to green innovation. All the barriers that are in the upper two quadrant of figure are selected for the second phase. All these barriers are having high driving power and

Table 8.8 Iteration 3 for level partition

Variable	Reachability set	Antecedent set	Intersection set	Level
BI01	1, 2, 3, 4, 7, 8, 10, 11	1, 2, 3, 4, 7, 8, 9, 10, 11	1, 2, 3, 4, 7, 8, 10, 11	III
BI02	1, 2, 3, 4, 7, 8, 11	1, 2, 3, 4, 7, 8, 9, 10, 11	1, 2, 3, 4, 7, 8, 11	III
BI03	1, 2, 3, 4, 7, 8, 9, 10, 11	1, 2, 3, 4, 7, 8, 9, 10, 11	1, 2, 3, 4, 7, 8, 9, 10, 11	III
BI04	1, 2, 3, 4, 7, 8, 10, 11	1, 2, 3, 4, 7, 8, 9, 10, 11	1, 2, 3, 4, 7, 8, 10, 11	III
BI07	1, 2, 3, 4, 7, 8, 11	1, 2, 3, 4, 7, 8, 9, 10, 11	1, 2, 3, 4, 7, 8, 11	III
BI08	1, 2, 3, 4, 7, 8, 11	1, 2, 3, 4, 7, 8, 9, 10, 11	1, 2, 3, 4, 7, 8, 11	III
BI09	1, 2, 3, 4, 7, 8, 9, 10, 11	3, 9, 10	3, 9, 10	
BI10	1, 2, 3, 4, 7, 8, 9, 10, 11	1, 3, 4, 9, 10, 11	1, 3, 4, 9, 10, 11	
BI11	1, 2, 3, 4, 7, 8, 10, 11	1, 2, 3, 4, 7, 8, 9, 10, 11	1, 2, 3, 4, 7, 8, 10, 11	III

Table 8.9 Iteration 4 for level partition

Variable	Reachability set	Antecedent set	Intersection set	Level
BI09	9, 10	9, 10	9, 10	IV
BI10	9, 10	9, 10,	9, 10	IV

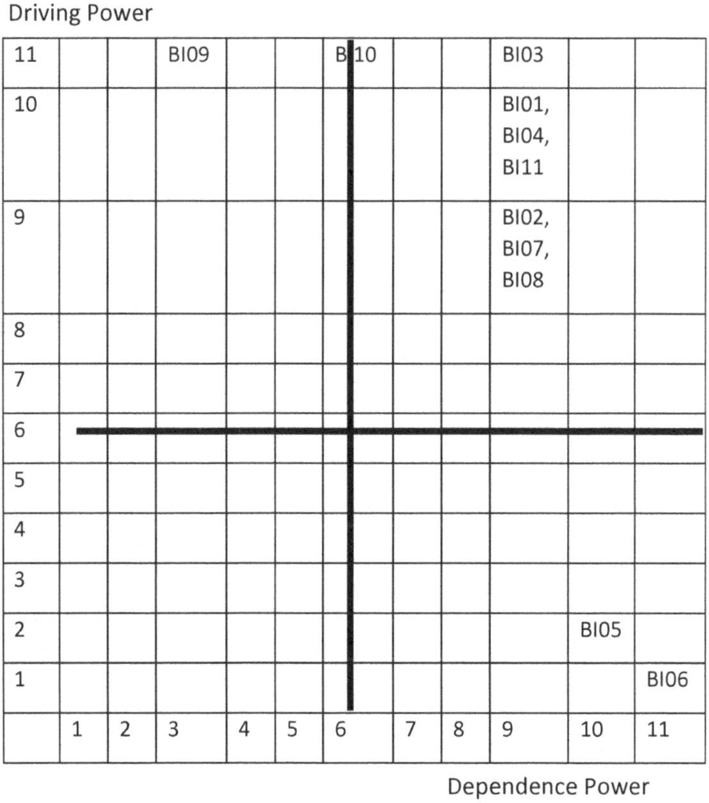

Fig. 8.2 MICMAC analysis on the basis of driving and dependence power

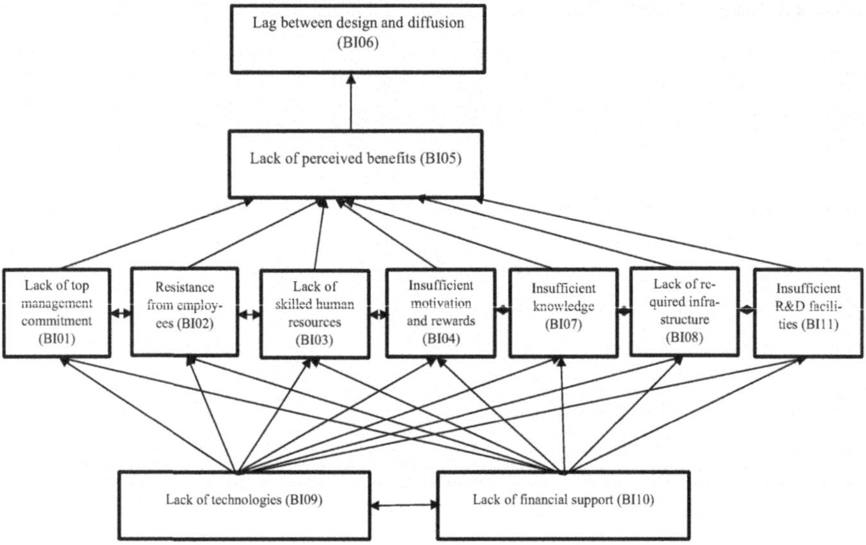

Fig. 8.3 ISM-based model for internal barriers to green innovation

are considered more important than other barriers. A total of nine barriers are selected for the second phase. The panel of four experts is asked to first identify best and worst barriers among the list of barriers and then give pairwise comparison ratings among best to others as well as others to best sing 9 point scale. These ratings are shown in Table 8.10 below.

After obtaining the pairwise ratings of best to others and others to worst barriers, next step is to obtain the weights of the barriers and hence their ranking. Using Eqs. 8.1 and 8.2 mentioned in Sect. 3.2, the weights and ranking of the barriers are obtained and are presented in Table 8.11.

According to best worst analysis, lack of financial resources (BI10) is the most important internal barrier to green innovation in manufacturing industries, followed by lack of technology (BI09) and lack of required infrastructure (BI08).

In the third phase of the research, the weights obtained through Best-Worst Method is used in VIKOR analysis to rank the best organization among four selected organizations on the basis of their ability to handle these internal barriers to green innovation. In this case, each expert is asked to individually rate the ability of four organizations to tackle these internal barriers to green innovation. The pairwise ratings of each of the organization w.r.t the listed nine barriers are obtained on a scale of 0–4 from each expert. Tables 8.12, 8.13, 8.14, 8.15, and 8.16 represent the ratings and average ratings given by four experts for different organizations.

Using the steps mentioned in Sect. 3.3, the ranking of each organization is obtained. The results (Table 8.17) show that the third organization is ranked first among all and other organizations can learn from the practices being followed at the top-ranked organization to incorporate green practices at their end.

Table 8.10 Pairwise comparison ratings of barriers to green innovation

Best to others	BI01	BI02	BI03	BI04	BI07	BI08	BI09	BI10	BI11
BI10	9	8	4	6	5	3	2	1	5
Others to the worst	BI01								
BI01	1								
BI02	2								
BI03	3								
BI04	2								
BI07	3								
BI08	4								
BI09	7								
BI10	9								
BI11	4								

Table 8.11 Weights and ranking of internal barriers to green innovation

	BI01	BI02	BI03	BI04	BI07	BI08	BI09	BI10	BI11
Weights	0.030	0.046	0.092	0.061	0.074	0.122	0.183	0.319	0.073
Ranking	9	8	4	7	5	3	2	1	6

Table 8.12 Rating of alternatives by Expert 1

	BI01	BI02	BI03	BI04	BI07	BI08	BI09	BI10	BI11
O1	3	2	2	3	2	3	3	3	2
O2	3	2	2	2	3	3	3	2	2
O3	4	2	3	3	3	2	2	2	1
O4	3	2	2	1	2	3	3	2	2

Table 8.13 Rating of alternatives by Expert 2

	BI01	BI02	BI03	BI04	BI07	BI08	BI09	BI10	BI11
O1	4	3	3	3	3	2	3	2	2
O2	3	2	3	2	2	4	3	2	1
O3	3	3	3	3	2	3	2	3	2
O4	2	2	1	1	1	3	3	3	3

Table 8.14 Rating of alternatives by Expert 3

	BI01	BI02	BI03	BI04	BI07	BI08	BI09	BI10	BI11
O1	3	3	4	4	3	3	2	2	3
O2	3	3	2	3	3	4	4	3	2
O3	3	4	3	3	4	3	2	3	1
O4	2	2	2	3	2	3	3	2	3

Table 8.15 Rating of alternatives by Expert 4

	BI01	BI02	BI03	BI04	BI07	BI08	BI09	BI10	BI11
O1	2	3	3	2	3	3	3	2	2
O2	3	3	3	2	3	3	3	2	2
O3	3	4	3	3	3	3	2	2	1
O4	3	2	1	3	1	3	3	2	2

Table 8.16 Average rating of the alternatives

	BI01	BI02	BI03	BI04	BI07	BI08	BI09	BI10	BI11
O1	3	2.75	3	3	2.75	2.75	2.75	2.25	2.25
O2	3	2.5	2.5	2.25	2.75	3.5	3.25	2.25	1.75
O3	3.25	3.25	3	3	3	2.75	2	2.5	1.25
O4	2.5	2	1.5	2	1.5	3	3	2.25	2.5
Max	3.25	3.25	3	3	3	3.5	3.25	2.5	2.5
Min	2.5	2	1.5	2	1.5	2.75	2	2.25	1.25

Table 8.17 S, R, and Q values for alternatives and ranking

	S	R	Q	Ranking
O1	0.570	0.319	0.765	3
O2	0.489	0.319	0.653	2
O3	0.379	0.183	0.000	1
O4	0.739	0.319	1.000	4
	$S^- = 0.739$	$R^- = 0.319$		
	$S^* = 0.379$	$R^* = 0.183$		

5 Discussions, Conclusions, and Scope of Future Work

The analysis of the case companies after the application of three-phase methodologies suggests that lack of technological abilities, lack of finance, and consequently lack of infrastructure are the major internal barriers to green innovation. Organizations are lacking on the front of availability of the latest technologies, like most organizations don't have any recycling facility at their end, and they are mostly dependent on external parties for recycling their products or by-products and often end up doing away with any recycling due to high cost of outsourcing. Thus organizations fail to recycle, reuse, or refurbish their products. This has great impact on environment as organizations are forced to use more and more of resources due to lack of technological facilities (Gupta and Barua 2017, 2018). Another barrier related to technologies is the inability of the organizations to use energy-efficient materials for their manufacturing; the organizations are not well equipped with the required technologies to process those materials, and they are often stuck on traditional raw materials which are harmful for the environment (Collins et al. 2007). Lack of finance is another major barrier to green innovation, and results of the case analysis also suggests that lack of finance in fact leads to all other barriers. Green innovations or any other innovation requires lot of research and development activities within the organizations; organizations are often short on funds and hence are unable to spend much on the research and development activities related to the greening of their activities and products (Kerr and Nanda 2015; Hall et al. 2011). Organizations often fail to get subsidies from government as well as subsidized loans from the financial institutions for their technology upgradation. Often the interest rates are so high that organizations avoid taking loans at such high interest rates and end up using the same old technologies over the years that are not suitable for producing environmental friendly products (Cecere et al. 2018). Also since organizations are already lagging in terms of technology, they are not able to dispose of the hazardous waste products due to high cost of outsourcing and hence act as deterrent to environmental management initiatives (Govindan et al. 2014). Another major barrier to green innovation is the lack of required infrastructure to carry out green innovations; the organizations don't have both latest technologies as well as infrastructure. Greening the processes requires processing of environmental friendly materials, disposing of hazardous materials, and recycling materials, but organizations are found to be lacking in having the required infrastructure to do this processing (Xia et al. 2019).

Economic development of any country depends upon the use and availability of natural resources. The more resources are available, the more of them can be utilized for the consumption-related needs and production to meet the demand of industry. Rapid industrialization has led to improvement in lifestyles and growth but at the same time has also caused replenishment of the resources along with deterioration of the environment. Environmental sustainability has gained attention in recent few years due to the havoc created by mass industrialization. These issues have forced managers to think of innovative ways to tackle environmental concerns and at the same time being sustainable so that future generations can also have sufficient availability of the resources for their survival. Green and sustainable innovation at the organizations is one such solution to achieve these goals. Managers need to find innovative ways of reduction in resource consumption like energy, material, and manpower. Also, they have to modify their current processes so that the whole production process emits the minimum hazardous substances in the environment. The current work focused on identifying internal barriers to green innovation. ISM and best worst analysis results can help managers of different organizations to identify the important barriers impacting green innovation and environmental goals.

The current research work has certain limitations, the current study is limited to manufacturing organizations only, and the results need to be compared with other sectors to further validate the results. Also, structural equation modeling can be applied on the hierarchal model obtained through ISM to statistically test the model.

References

Ahmadi, H. B., Kusi-Sarpong, S., & Rezaei, J. (2017). Assessing the social sustainability of supply chains using Best Worst Method. *Resources, Conservation and Recycling, 126*, 99–106.

Al-Muftah, H., Weerakkody, V., Rana, N. P., Sivarajah, U., & Irani, Z. (2018). Factors influencing e-diplomacy implementation: Exploring causal relationships using interpretive structural modelling. *Government Information Quarterly, 35*(3), 502–514.

Cecere, G., Corrocher, N., & Mancusi, M. L. (2018). Financial constraints and public funding of eco-innovation: Empirical evidence from European SMEs. *Small Business Economics, 54*, 285–302.

Chang, C. H. (2011). The influence of corporate environmental ethics on competitive advantage: the mediation role of green innovation. *Journal of Business Ethics, 104*(3), 361–370.

Chen, Y. S., Lai, S. B., & Wen, C. T. (2006). The influence of green innovation performance on corporate advantage in Taiwan. *Journal of Business Ethics, 67*(4), 331–339.

Collins, E., Lawrence, S., Pavlovich, K., & Ryan, C. (2007). Business networks and the uptake of sustainability practices: the case of New Zealand. *Journal of Cleaner Production, 15*(8-9), 729–740.

De Jesus, A., & Mendonça, S. (2018). Lost in transition? Drivers and barriers in the eco-innovation road to the circular economy. *Ecological Economics, 145*, 75–89.

Del Río, P., Carrillo-Hermosilla, J., & Könnölä, T. (2010). Policy strategies to promote eco-innovation: An integrated framework. *Journal of Industrial Ecology, 14*(4), 541–557.

Dubey, R., Gunasekaran, A., & Ali, S. S. (2015). Exploring the relationship between leadership, operational practices, institutional pressures and environmental performance: A framework for green supply chain. *International Journal of Production Economics, 160*, 120–132.

Dubey, R., Gunasekaran, A., Childe, S. J., Papadopoulos, T., Hazen, B., Giannakis, M., & Roubaud, D. (2017). Examining the effect of external pressures and organizational culture on shaping performance measurement systems (PMS) for sustainability benchmarking: Some empirical findings. *International Journal of Production Economics, 193*, 63–76.

Eiadat, Y., Kelly, A., Roche, F., & Eyadat, H. (2008). Green and competitive? An empirical test of the mediating role of environmental innovation strategy. *Journal of World Business, 43*(2), 131–145.

Fernando, Y., Jabbour, C. J. C., & Wah, W. X. (2019). Pursuing green growth in technology firms through the connections between environmental innovation and sustainable business performance: does service capability matter? *Resources, Conservation and Recycling, 141*, 8–20.

Govindan, K., Kaliyan, M., Kannan, D., & Haq, A. N. (2014). Barriers analysis for green supply chain management implementation in Indian industries using analytic hierarchy process. *International Journal of Production Economics, 147*, 555–568.

Gupta, H. (2018). Evaluating service quality of airline industry using hybrid best worst method and VIKOR. *Journal of Air Transport Management, 68*, 35–47.

Gupta, H., & Barua, M. K. (2016). Identifying enablers of technological innovation for Indian MSMEs using best–worst multi criteria decision making method. *Technological Forecasting and Social Change, 107*, 69–79.

Gupta, H., & Barua, M. K. (2017). Supplier selection among SMEs on the basis of their green innovation ability using BWM and fuzzy TOPSIS. *Journal of Cleaner Production, 152*, 242–258.

Gupta, H., & Barua, M. K. (2018). A framework to overcome barriers to green innovation in SMEs using BWM and Fuzzy TOPSIS. *Science of the Total Environment, 633*, 122–139.

Hafezalkotob, A., Hafezalkotob, A., Liao, H., & Herrera, F. (2019). An overview of MULTIMOORA for multi-criteria decision-making: Theory, developments, applications, and challenges. *Information Fusion, 51*, 145–177.

Hall, J., Matos, S., Silvestre, B., & Martin, M. (2011). Managing technological and social uncertainties of innovation: The evolution of Brazilian energy and agriculture. *Technological Forecasting and Social Change, 78*(7), 1147–1157.

Hermann, R. R., Mosgaard, M. A., & Kerndrup, S. (2016). The function of intermediaries in collaborative innovation processes: Retrofitting a Danish small island ferry with green technology. *International Journal of Innovation and Sustainable Development, 10*, 361.
Hughes, D. L., Rana, N. P., & Dwivedi, Y. K. (2019). Elucidation of IS project success factors: An interpretive structural modelling approach. *Annals of Operations Research, 285*, 35–66.
Kerr, W. R., & Nanda, R. (2015). Financing innovation. *Annual Review of Financial Economics, 7*, 445–462.
Liao, W., & Wang, T. (2018). Promoting green and sustainability: A multi-objective optimization method for the job-shop scheduling problem. *Sustainability, 10*(11), 4205.
Lin, C. Y., & Ho, Y. H. (2008). An empirical study on logistics service providers' intention to adopt green innovations. *Journal of Technology Management & Innovation, 3*(1), 17–26.
Madrid-Guijarro, A., Garcia, D., & Van Auken, H. (2009). Barriers to innovation among Spanish manufacturing SMEs. *Journal of Small Business Management, 47*(4), 465–488.
Malek, J., & Desai, T. N. (2019). Interpretive structural modelling based analysis of sustainable manufacturing enablers. *Journal of Cleaner Production, 238*, 117996.
Mangla, S. K., Govindan, K., & Luthra, S. (2017). Prioritizing the barriers to achieve sustainable consumption and production trends in supply chains using fuzzy Analytical Hierarchy Process. *Journal of cleaner production, 151*, 509–525.
Opricovic, S. (1998). Multicriteria optimization of civil engineering systems. *Faculty of Civil Engineering, Belgrade, 2*(1), 5–21.
Pawanchik, A., & Sulaiman, S. (2010). *In search of InnovAsian: the Malaysian innovation climate report 2010*. Alpha Catalyst Consulting.
Perron, G. M. (2005). *Barriers to environmental performance improvements in Canadian SMEs*. Canada: Dalhousie University.
Porter, M. E., & Van der Linde, C. (1995). Green and competitive: ending the stalemate. *Harvard Business Review, 73*(5), 120–134.
Rezaei, J. (2015). Best-worst multi-criteria decision-making method. *Omega, 53*, 49–57.
Rezaei, J. (2016). Best-worst multi-criteria decision-making method: Some properties and a linear model. *Omega, 64*, 126–130.
Rezaei, J., Nispeling, T., Sarkis, J., & Tavasszy, L. (2016). A supplier selection life cycle approach integrating traditional and environmental criteria using the best worst method. *Journal of Cleaner Production, 135*, 577–588.
Saunila, M., Ukko, J., & Rantala, T. (2018). Sustainability as a driver of green innovation investment and exploitation. *Journal of cleaner production, 179*, 631–641.
Song, M., & Wang, S. (2018). Market competition, green technology progress and comparative advantages in China. *Management Decision, 56*(1), 188–203.
Song, M., Peng, J., Wang, J., & Dong, L. (2018). Better resource management: An improved resource and environmental efficiency evaluation approach that considers undesirable outputs. *Resources, Conservation and Recycling, 128*, 197–205.
Tam, V. W., Tam, C. M., Shen, L. Y., Zeng, S. X., & Ho, C. M. (2006). Environmental performance assessment: perceptions of project managers on the relationship between operational and environmental performance indicators. *Construction Management and Economics, 24*(3), 287–299.
Walker, H., Di Sisto, L., & McBain, D. (2008). Drivers and barriers to environmental supply chain management practices: Lessons from the public and private sectors. *Journal of purchasing and supply management, 14*(1), 69–85.
Warfield, J. N. (1974). Developing interconnection matrices in structural modeling. *IEEE Transactions on Systems, Man, and Cybernetics*, (1), 81–87.
Xia, D., Zhang, M., Yu, Q., & Tu, Y. (2019). Developing a framework to identify barriers of Green technology adoption for enterprises. *Resources, Conservation and Recycling, 143*, 99–110.
Zhao, H., Guo, S., & Zhao, H. (2018). Comprehensive benefit evaluation of eco-industrial parks by employing the best-worst method based on circular economy and sustainability. *Environment, Development and Sustainability, 20*(3), 1229–1253.

Zhe, L., Yong, G., Hung-Suck, P., Huijuan, D., Liang, D., & Tsuyoshi, F. (2016). An emergy-based hybrid method for assessing industrial symbiosis of an industrial park. *Journal of Cleaner Production, 114*, 132–140.

Zhu, Q., & Sarkis, J. (2007). The moderating effects of institutional pressures on emergent green supply chain practices and performance. *International Journal of Production Research, 45* (18-19), 4333–4355.

Chapter 9
Strategic and Managerial Decision-Making for Sustainable Management: Factors and Remedies for Information Overload

Geerten van de Kaa

Abstract In the past century, the amount of information that is generated has increased at an exponential rate. As a result of this information saturation process or 'information overload', people cannot keep pace with the information flowing towards them. Organizations can take steps towards sustainability by improving the health conditions in their organizations and the products they offer. One way to do so is by trying to reduce information overload. Besides improving health conditions, decreasing information overload also positively affects the quality of organizational-level decision-making. This is a conceptual chapter in which propositions are built. The determinants of information overload and the effect on decision quality are described, and an extensive literature review is conducted. The chapter draws from decision theory and the theory of human information processing and focuses on the individual characteristics of decision-makers and the quality of the information provided by the information system. Seven propositions are formulated and several remedies for information overload are discussed. The chapter arrives at two determinants of an individual's degree of information overload experienced (characteristics of the decision-maker and the information system), and it is argued that information overload has a negative influence on the quality of decisions. These concepts are combined in one model. The conceptual model is an attempt to provide a better theoretical understanding of the concept of information overload. Finally, contributions, limitations and areas for future research are discussed.

Keywords Information overload · Sustainability · Health conditions · Decision quality

---

G. van de Kaa (✉)
Faculty of Technology, Policy, and Management, Delft University of Technology, Delft, The Netherlands
e-mail: g.vandekaa@tudelft.nl

J. Rezaei (ed.), *Strategic Decision Making for Sustainable Management of Industrial Networks*, Greening of Industry Networks Studies 8,
https://doi.org/10.1007/978-3-030-55385-2_9

1 Introduction

On a daily basis, people spend a considerably amount of time on such tasks as checking emails, typing tweets and checking and updating Facebook and LinkedIn. In the beginning of the twentieth century, the main characteristic of information was its scarcity. This situation has gradually been changing, bringing more information and increasing its quality and its diversity. The introduction and evolvement of information and communication technologies such as personal computers, the Internet and smart phones have driven humanity towards the digital economy which is characterized by an abundance of information.

However, the huge growth of information has been accompanied by a decline in its marginal value of meaning comparable to the inflation of currency; the more there is of it, the less it seems to 'buy' in meaning (Klapp 1986). As a result of this information saturation process or 'information overload' (IO), people cannot keep pace with the information flowing towards them. The phenomenon of IO, which is a result of the abundance of information provided by ICTs, negatively impacts various social communities and society at large. Strategic and managerial decisions should be taken to decrease IO and thereby improve the health conditions in organizations and increase sustainability.

The topic of IO has been studied from multiple angles including organization, marketing, accounting and management information systems by both social and technical scientists. However, few researchers have done interdisciplinary research into the topic of IO (Eppler and Mengis 2004). This research combines the study of information technology in its pure terms with the study of limitations that are imposed on the decision-maker.

The objective of this research is to achieve a better theoretical understanding of the concept of IO, by studying the interaction between the decision-maker and the information provided by the information system. I aim to conceptualize the relation between the cognitive and personality characteristics of decision-makers and the quality of the information provided by the information system and their impact on the degree of IO decision-makers' experience. The practical significance of this research is that it can give practitioners a better understanding of IO.

We approach the decision-maker as a certain combination of cognitive and personality characteristics that determine the efficiency of information processing and, as a result, the decision-making outcome. We adopt the theoretical framework proposed by the decision-making school claiming that technology is designed in order to overcome human weaknesses (DeSanctis and Poole 1994). However, we step aside from the major implication that is often driven from the decision-making theoretical perspective. While the representatives of this school tend to agree that factual technological inefficiencies result from errors in systems design, development or implementation phases, we focus on the interaction between the decision-maker and the information system and attempt to provide a logical explanation to the paradox why information systems that are initially designed to improve human information processing (HIP) often add to individual IO and reduce decision quality.

The goal is to achieve a more comprehensive understanding of the relationship between the personality characteristics of the decision-maker, the information system and the degree of individual IO.

We develop a conceptual model to investigate managerial decision-making under IO and examine the trade-off between information processing capacity and information processing requirements. Considering the abundance of research done in this area, we impose a number of restrictions, thus making the research feasible. In our terms, information processing capacity is determined by the cognitive and personality characteristics of the decision-maker. Information processing requirements that are imposed by a task are taken as given. However, the decision-maker is free in choosing information acquiring and information processing behaviour. The information system, the main source of the internal corporate data, is assumed to be the most influential determinant of the organizational context. The efficiency of the information system and the quality of the information output thus affects individual decision-making in a significant way, reinforcing or eliminating IO.

Indeed, one particular way to both increase the quality of organizational-level decision-making and to increase sustainability within organizations and improve health conditions in organizations is to attempt to reduce information overload. At the end of this chapter, we will discuss several countermeasures for information overload.

2 Theory and Propositions

2.1 The Relation Between Information Overload, Sustainability and Organizational-Level Decision-Making Quality

The rational decision-maker systematically looks for the decision that will maximize results. The obtained solutions must be as close as possible to the optimal solution. We assume that managers have perfect information, have unlimited information processing capacity, will end up with the optimal solution and be able to predict the consequences of actions. Simon (1955) presented the behavioural decision model . It studies the processes that managers actually use when selecting from among alternatives. In the behavioural decision model, decision-makers are affected by bounded rationality. This refers to the idea that people are limited in the processing of information by constraints including their information processing capabilities. Due to these limitations, decision-makers are forced to make decisions that are just good enough or decisions that are satisficing. So in the rational decision model, people look for the optimal solution, whereas in the behavioural decision model, people look for the most satisficing solution. In this chapter, we position ourselves in the behavioural decision model and argue that IO is another important limitation in the decision-making process.

There are a number of studies which have studied the negative relationship between information overload and quality of organizational-level decision-making (Good 1958; Schroder et al. 1967). For example, in 1996 a survey of 1313 junior, middle and senior managers in the UK, the USA, Australia, Hong Kong and Singapore was performed, and it showed that 43% of managers think important decisions are delayed due to information overload. The quality of decisions greatly decreases when decisions are delayed because most of the decisions have to be made at one particular moment in time. A delay in decision-making means a decrease in the overall quality of the decisions. Information overload also causes the response rate to decrease and the performance to degrade (Good 1958). There are also various studies that stress the negative relationship between information overload and organizational sustainability in terms of health conditions of employees. For example, Swar (Bawden and Robinson 2008; Kim et al. 2007; Misra and Stokols 2012; Swar et al. 2017). This is because the information that is available to the person is more than what can be managed by that person leading to coping strategies that are not effective and greater than what the information seeker can manage effectively leading to ineffective coping strategies and indications of psychological illness (Hall and Walton 2004). This paper focuses on specific aspects of the decision-makers and the information system that determine the perceived information overload.

2.2 Human Information Processing

When people make a decision, they process information. Decision-making is a specific form of HIP. Therefore, we will use HIP frameworks as a guideline for our research on the decision-maker. The information processing capacity of decision-makers partly determines how much IO they experience. Rumelhart (1977) distinguished five main stages of HIP: sensing, recognizing patterns, understanding language, remembering and reasoning. Sensing is defined as the point at which fluctuations of energy in the environment are translated into fluctuations of neural activity. The better a person is at sensing, the more fluctuations of energy can be translated into fluctuations of neural activity. Recognizing patterns is the second stage of HIP. Several pieces of data can be presented in different ways but they have the same meaning. A person has to discover a pattern between these pieces of data. The third stage is the understanding of language. The processing system combines information from the sense organs with the knowledge of the language to determine the interpretation of speech or written language. The fourth stage of HIP is remembering. The memory of an individual can best be compared to the memory of a computer. The internal memory is easily accessible, whereas the external memory is more difficult to access. In the human mind, the internal memory is the part of the memory that is activated. It could be information that we have just received or memory that we have just retrieved from our external memory. In this chapter the focus lies on working memory. There are limits to how much information can be activated at the same time. This differs from individual to individual. The most important difficulty in remembering is the problem of retrieval difficulty and not the problem of information loss. Because the secondary memory is well structured, information can easily be retrieved from it. The last stage of HIP is reasoning. Rumelhart (1977) uses the term reasoning to denote 'those processes of information retrieval that rely for success on the structure, as opposed to the content, of organised memory'.

The working memory and the attention capacity of an individual play an important role in the processing of information. We assume these two cognitive characteristics partly determine whether a decision-maker experiences IO. Basically, HIP is about accepting input in the form of information and producing output in the form of decisions. When the human capacity for processing information is surpassed, IO can occur. The capacity to accept input and produce outputs is determined by the decision-maker's cognitive and personality characteristics.

2.3 Decision-Maker's Cognitive Characteristics

Cognitive characteristics refer to the characteristics of the decision-maker which are connected with the human brain and in a way determined by nature. The focus lies on two concepts which are related to the cognitive characteristics of the decision-maker: working memory and attention capacity.

Working memory is that part of the memory that keeps information active while we are using it or until we use it (Just and Carpenter 1992). The better a person is at comprehension, the better his/her working memory is. When the person is good at comprehending difficult tasks, this person can distinguish irrelevant information from relevant information better than people who are worse at comprehending difficult tasks. The faster memory is retrieved from the working memory, the faster new information can be put in working memory. This means the better the working memory of an individual, the more information he/she can process in a given time, and thus the less IO he/she will perceive.

Proposition 1 The more working memory a person has, the less information overload he/she will perceive positively affecting organizational health.

According to Conway and Engle (1996), working memory is greatly linked to the attention capacity of an individual which can be defined, according to cognitive psychologists, as the ability to concentrate. The limitations of human attention represent one of the bottlenecks in HIP. The concept becomes necessary because we do not process all stimuli that impinge upon us. We often neglect to attend to stimuli because we are distracted. People have so many tasks to perform that some of the tasks are neglected. Attentional resources can be thought of as a pool from which all tasks and mental activities are drawn. One important property of attentional resources is the fact that these resources are limited. The fact that they are limited means they have to be divided among different tasks on which one is working. When a decision-maker has a high attention capacity, he/she can divide more attention, and this would mean he/she can concentrate on more tasks. This means he/she will experience less IO than the person who can divide less attention among the tasks. With the increase of information load, the demand on attentional resources also increases. When a person has a low attention capacity and information load is high, this person can get overload with information because he/she has to concentrate on too much information originating from the many tasks. In this line of reasoning, attention capacity determines the level of IO which a person experiences. The more attention capacity a person has, the more information he/she can process, and thus the less overloaded he/she will be with information.

Proposition 2 The more attention capacity a person has, the less information overload he/she will perceive positively affecting organizational health.

It is assumed that both the working memory and the attention capacity have a negative impact on individual IO. So, the higher the working memory capacity or attention capacity, the lower IO will become. We propose attention capacity has a more positive impact on individual IO than working memory capacity. When a person has a high attention capacity, he/she can concentrate longer but not faster. People having a high working memory can process the excess of information faster because they can comprehend complex tasks faster. Given the fact that there is a time constraint, decision-makers having a higher working memory compared to attention capacity will experience less IO in a given time constraint than people having a higher attention capacity compared to their working memory capacity. This results in the following proposition:

Proposition 3 Decision-makers having a higher working memory compared to attention capacity will experience less information overload than people having a higher attention capacity compared to their working memory capacity.

2.4 Personality Characteristics

Social scientists assume that behaviour is a function of personality variables and that to understand human behaviour, it is important to look at the personality characteristics of the human being. Three personality characteristics seem to be important: the amount of information used, the decision focus and the attitude towards ambiguity. These concepts are developed over a lifetime and can also change during one's lifetime. We propose that these characteristics have an impact on individual IO.

Information needs rise when people experience cognitive gaps. To bridge these gaps, good, accessible information has to be sought. People who take decisions based on a limited amount of information want just enough data to make an adequate decision and want to save time. However, people who take decisions based on a maximum amount of information want to process as much information as possible before making a decision, and time is less important for them. There are several ways to process information. The most obvious way is reading an information source. Let us assume that a decision-maker has to write a report based on a certain information source. A person who prefers to read an article that contains a high amount of information clearly has a preference for high quantities of information. Another measure of preference for amount of information used for decision-making is by looking at how much information a person writes. Writing is, in fact, a form of accessing information. For instance, when people write a report, they retrieve information from both their long-term and their short-term memory. Thus, people who prefer to write a detailed rather than a brief report have to retrieve more information from their short-term and long-term memory and will therefore use a higher amount of information. We think the more information people want to use, the more likely they will be overloaded with information because they cannot process the information. People who prefer a limited amount of information but who are confronted with large amounts of information will not process the excess information and thus will not suffer from information overload. Thus, we hypothesize:

Proposition 4 The amount of information used has a positive impact on IO. The greater the amount of information people are willing to process, the more IO they are likely to experience negatively affecting organizational health.

The second aspect by which we can characterize people is their degree of focus in the decision-making process. Decision-makers who focus on only one solution will interpret the information leading to only one decision, whereas those who focus on multiple solutions will interpret the information leading to multiple decisions. Decision focus can be measured by looking at how people view the decision problem. They can tackle the problem as a whole, or they can focus on details. If decision-makers see the problem as a whole, they tend to focus on multiple solutions, whereas if they concentrate on details, they tend to focus on one solution. Decision focus can also be measured by looking at the way decision-makers sum up the information before making up their mind. This could include as many alternatives as possible, several alternatives with contradicting scenarios, several alternatives that reconfirm and add to one another and finally the single alternative that is the most convincing in their perception. We think that when decision-makers prefer to focus on multiple solutions/alternatives rather than one solution/alternative, they are more likely to experience IO because they have to spend time evaluating many solutions/alternatives. This results in the following hypothesis:

Proposition 5 Decision focus has a negative impact on IO. The more focused people are on one solution, the less IO they are likely to experience positively affecting organizational health.

The third personality characteristic is attitude towards ambiguity. Ambiguity is a situation where there is doubt or uncertainty as regards to the interpretation of a piece of data. Information can quickly become ambiguous if there are large quantities of it. People can have a low or a high tolerance for ambiguity. Decision-makers with low ambiguity tolerance who are confronted with large amounts of ambiguous information will be less certain about their decisions and will thus want to process more information. As a result, they will feel information overload more quickly than people who have a high tolerance for ambiguity. However, people with high ambiguity tolerance can perform better when confronted with an increasing amount of information, and their level of IO will be lower than people who have a low tolerance for ambiguity. Therefore, tolerance for ambiguity can determine perceived IO. This results in the following proposition:

Proposition 6 Tolerance for ambiguity has a negative impact on IO. The more tolerant people are towards ambiguity, the less IO they are likely to experience positively affecting organizational health.

2.5 Information Quality

Information quality is largely determined by the availability, accuracy and understandability of information. Important documents may be available but inaccessible because they are stored in numerous places and various conflicting standards may be used. This lack of a cohesive and standard system to make important information accessible results in IO. If relevant information were easily available, the quality of information would increase and perceived IO would decrease. We propose that if information systems provide more relevant information, decision-makers would not have to spend time evaluating irrelevant information, thus reducing perceived IO.

Proposition 7 The availability of information provided by the information system has a negative impact on IO. The more people perceive the information provided by the information system to be relevant, the less IO people are likely to experience positively affecting organizational health.

Accuracy of information is a second determinant of information quality. The decision-maker evaluates the accuracy of the information. When users come across information they think might be inaccurate, they have to check its validity. The time this takes cannot be spent processing accurate information. When there is much inaccurate information, much information must be evaluated without a purpose. This means time is wasted. Correspondingly, perceived IO will be higher when information accuracy is lower. Therefore, we posit the following proposition:

Proposition 8 The accuracy of information provided by the information system has a negative impact on IO. The more people perceive the information provided by the information system to be accurate, the less IO people are likely to experience.

When people do not understand the information generated by the information system, they are confronted with low-quality information. Information becomes less understandable when it is duplicate, unclear, false or anonymous. When information is duplicate, the user gets confused and cannot understand the text. Unclear information can occur due to, for example, a poor-quality interface. When people think the information provided is true, but in fact it is false, they are misled. When a source of information is anonymous, people may be reluctant to believe or trust the information. Every author has a different style and approaches a topic in a different manner. Thus, when the author is not known, a person accessing the information source via the information system may not understand the text. We propose IO perceived is lower when information is understandable and is presented in a precise, complete and clear way.

Proposition 9 The understandability of information provided by the information system has a negative impact on IO. The more people perceive the information provided by the information system to be understandable, the less IO they are likely to experience positively affecting organizational health.

3 Countermeasures to Reduce IO and Improve Organizational Health Conditions

We have argued that, among others, individual characteristics of decision-makers determine the information overload perceived. Therefore, countermeasures to reduce information overload can partly be pointed towards changing these characteristics. Changing one's individual characteristics is very difficult though and perhaps even impossible. There exist genes which are linked to behaviour, including thrill seeking and anxiety. In fact, there is clear evidence that suggests that genetic factors determine the development and functioning of personality (Jang 2001). Then, because one cannot chance one's genes, one can also not change ones individual characteristics. However, there do exist possibilities to decrease the information overload at the individual level. The first method to decrease the information overload at the individual level is by improving a person's cognitive functions. This can be reached by training the working memory capacity of the individual decision-maker (by, e.g. mnemonics). People could also train in speed reading or text scanning where text is quickly read and absorbed. The more people read texts and write reports, the more they will learn the ability to scan texts. When the personal management of information is optimized, the chances that information overload will occur will be lower. A straightforward example is the desktop in the decision-maker's office. When it is clean, information can easier be found. Another example is the personal management of files on the computer. Microsoft Windows makes use of a 'my documents' folder. Most people using the text processor word will save important documents in this folder and do not have folders in the 'my documents' folder. Email overload is a problem which is affecting managers severely. When looking at the problem of email overload, there are several possibilities with respect to the management of emails. Whittaker and Sidner (1996) have done empirical research into the topic of email overload and have identified three different user strategies, based on two criteria: whether or not users use folders and how users clean up there inbox. It resulted in three possible strategies for email management: no filers, frequent filers and spring cleaners. No filers are people who do not use folders, frequent filers are people how use folders and clean up their inbox every day and spring cleaners are people who use folders and clean up their inbox periodically. The inboxes of the no filers were definitely overloaded with a mean of 3093.5 emails. The strategy of the no filers to reduce the size of the email box was to delete large quantities of emails periodically. An interesting result of the study was that four of the six no filers were managers. The email boxes of the frequent filers were relatively small in size with a mean of 43.4 emails and the frequent filers were thus not overloaded with emails. Again an interesting result was that of the five frequent filers, only one was a manager. The spring cleaners had large overloaded email boxes with a mean of 1492.3. Most of the spring cleaners did not use the folders at all resulting in a large email box. Again four of the seven spring cleaners were managers. One can conclude that people who use folders and who clean up their emails on a daily basis are the least overloaded with respect to emails. However, as concluded by the study, managers do not use this strategy. A possibility to reduce the amount of emails which one received is to instruct the email program to filter out email which were sent as a 'cc'. Our recommendation to managers thus will be to structure their personal management of incoming information to reduce the chances of being overload with information. Another way of decreasing information overload is by properly communicating with others. One way of doing so is by adhering to the rules of netiquette. A simple example is to use proper subject lines when sending emails.

Countermeasures to decrease information overload can also be pointed to increasing the quality of information which is provided by the information system. In the literature there are numerous techniques suggested to increase the quality of information. Most of the techniques make use of the concept of an agent; therefore we will first concentrate on this concept.

To understand intelligent search agents, we will first give a formal definition of an agent. According to Woolridge and Jennings, 'an agent is a (computer) system that is situated in some environment and that is capable of autonomous action in this environment to meet its objectives' (Woolridge and Jennings 1995). Examples are robots, softbots and software demons. There are three main characteristics of the intelligent agent that make it flexible. Intelligent agents are reactive, proactive and have a social ability. Reactive means that intelligent agents can perceive their environment and can respond to changes in the environment to satisfy its objectives. Proactive means that intelligent agents can take the initiative in order to satisfy its objectives. The fact that the intelligent agent has a social ability means that he can communicate and cooperate with other agents to satisfy its objectives. According to Russell and Norvig (1995), "an agent is anything that can be viewed as perceiving its environment through sensors and acting upon that environment through effectors". Due to the three characteristics of the agent mentioned earlier, the agent will react to information, and it will, for instance, look for irrelevant information to filter out. In the process of doing so, it will possibly communicate with other agents so that it knows what exactly is irrelevant data and what exactly is relevant data. Lastly the agents will communicate the relevant information to other agents. In this respect it is good to note that agents can also be human individuals such as, in our case, decision-makers. This means agents can filter our irrelevant data, they can make the data more accurate and they can present the data in a more understandable format. There are several agents which have been developed. The intelligent filtering agent has as its primary objective to filter out irrelevant data, and the intelligent interface agent has as its primary objective to present data in a more understandable format. We will now concentrate on these two agents.

The volume of electronic mail is growing, and some decision-makers have to spend almost half an hour every day replying to electronic mails. Mailing lists, bulletin boards, asynchronous computer conferences and newsgroups send messages to many recipients; part of these messages are unwanted because users are not subscribed anymore. An intelligent filtering agent can filter out irrelevant emails and only present the relevant emails to the decision-maker. By applying information filtering techniques, irrelevant data can be reduced and information can become more relevant. Artificial intelligence studies how this filtering can be achieved. One of the many techniques which come from the field of artificial intelligence is the artificial neural network. These networks are based on the way the brain processes information. The network is composed of a large number of highly interconnected information processing elements that can be compared to the neurons we have in our brain. These so-called perceptrons are tied together with weighted connections that can be compared with the synapses that we have in our brain.

Another type of intelligent agent is the intelligent interface agent. However, when looking at intelligent interfaces, we still have a long way to go before these interfaces can be used to make the information which is presented to the decision-maker more clear. For now it is best to improve the interface by looking at the design of the interface. The information system presents the information through an interface. In the beginning of the computer era, interfaces were command line-based, the user would have to type in a command and the computer performs an action. With the introduction of the mouse, it became possible to use graphical user interfaces. Nowadays almost every computer makes use of the graphical user interface. According to scientific research, humans understand images and visual information better than plain text. When focusing on the presentation of data as a determinant of information quality, it is good to study human-computer interaction, also known as ergonomics.

Another type of interface is a voice recognition interface. Information can be presented to the user via a voice, the user does not have to read the information on a computer screen, but the user will hear the information wherever his location may be. Although this type of interface is still in the early phases of development, we will concentrate on it because we think it will be one of the most important techniques which can help increase all the three concepts of information quality as discussed in Sect. 2. Currently, with voice recognition software, the user has to go through a number of exercises in which he has to 'train' the voice recognition software so that it will understand the speaker's voice. If there would be a possibility to really speak to the computer without pauses between the words spoken and without the limitations of a specific library of words, the ultimate goal has been reached. This type of technologies is called natural speech input. The problem lies in making the conversation between the user and the computer as human-like as possible. When this technology is sufficiently mature`, decision-makers can ask which information they want to have, and they are presented with the relevant, accurate information in an understandably way. This increases information quality and thus decreases information overload.

Theoretically it is possible to create a software agent that registers what a manager does every day in terms of appointments, what emails he processes, which documents he works on and what he has to do the next day. In doing so the agent learns a pattern, and based upon this pattern, the software agent understands more which type of information the manager prefers and how he would like it to be presented. Information gets more relevant, accurate and understandable in this way. This is an example of a product which tries to solve the problem of organizational information overload.

4 Conclusion

In this chapter we have focused on two determinants of an individual's degree of IO experienced, and we have argued that IO has a negative influence on the quality of decisions. In Fig. 9.1 we combine these concepts in one model. The conceptual model is an attempt to provide a better theoretical understanding of the concept of IO.

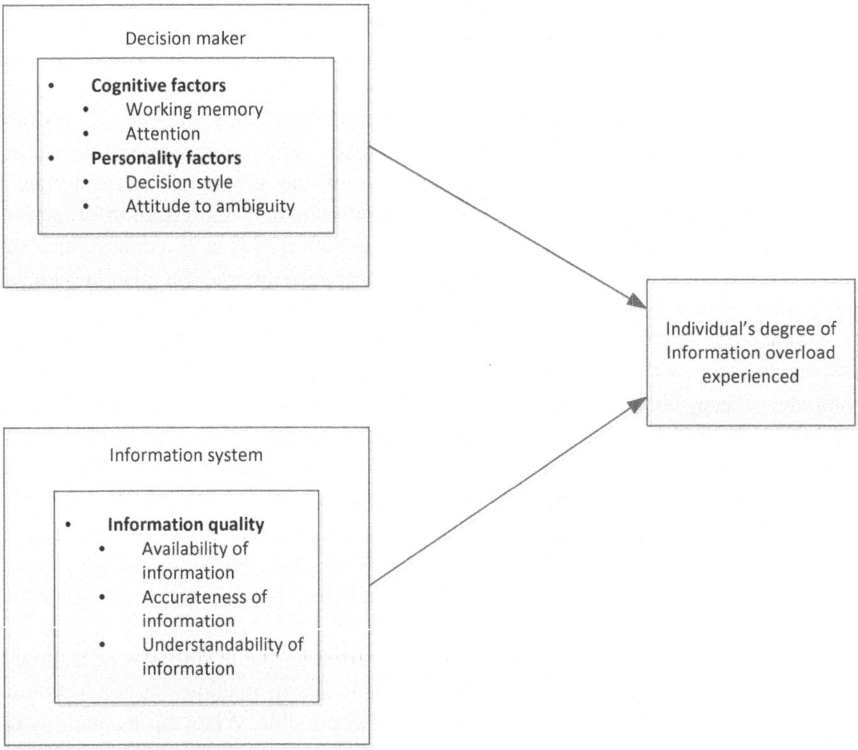

Fig. 9.1 Conceptual model

The conceptual model can be used for further research into the topic of IO. Future research can attempt to test the propositions put forth in this chapter. We argue that information overload can partly be reduced by a change in a person's personal characteristics. Also, information overload can partly be reduced by presenting the information provided by the information system in a clear and unambiguous way. The insights from this study can be useful for information intensive organizations in developing countermeasures for information overload and thereby increasing overall health conditions with the organization (and thus increasing sustainability). These organizations could also use the outcomes of this research to advise other firms about dealing with information overload.

In this chapter, we focused on two determinants of an individual's perceived degree of information overload. We examined several cognitive and personality traits of decision-makers and studied three concepts of information quality provided by information systems. However, for both determinants other characteristics could have an effect on the perceived IO. In future research these other characteristics could be studied.

Recommendations go out to the IT departments of organizations. These departments can reduce the information overload significantly by improving the relevance, accurateness and understandability of information. Information technology can be applied to reduce the amount of irrelevant information and improve the accurateness of information. Applying intelligent agents from the field of artificial intelligence could make information systems 'smarter' which would reduce information overload.

References

Bawden, D., & Robinson, L. (2008). The dark side of information: Overload, anxiety and other paradoxes and pathologies. *Journal of Information Science, 35*, 180–191.

Conway, A. R. A., & Engle. (1996). Individual differences in working memory capacity: More evidence for a general capacity theory. *Memory, 4*(6), 577–590.

DeSanctis, G., & Poole, M. S. (1994). Capturing the complexity in advanced technology use: Adaptive structuration theory. *Organization Science, 5*(2), 121–147.

Eppler, M., & Mengis, J. (2004). The concept of information overload: A review of literature from organization science, accounting, marketing, MIS, and related disciplines. *Information Society, 20*(5), 325–344.

Good, H. H. (1958). Greenhouses of science for management. *Management Science, 4*(4), 365–381.

Hall, A., & Walton, G. (2004). Information overload within the health care system: A literature review. *Health Information Libraries Journal, 21*, 102–108.

Jang, K. L. (2001). Behavioural-genetic perspectives on personality function. *Canadian Journal of Psychiatry, 46*(3), 234–244.

Just, M. A., & Carpenter, P. A. (1992). A capacity theory of comprehension: Individual differences in working memory. *Psychological Review, 99*(1), 122–149.

Kim, K., Lustria, M. L. A., & Burke, D. (2007). Predictors of cancer information overload: Findings from a national survey. *Information Research, 12*, 1–29.

Klapp, O. E. (1986). *Overload and boredom: Essays on the quality of life in the information society*. Westport: Greenwood Publishing Group Inc.

Misra, S., & Stokols, D. (2012). Psychological and health outcomes of perceived in-formation overload. *Environment and Behavior, 44*, 737–759.
Rumelhart, D. E. (1977). *Introduction to human information processing*. New York: Wiley.
Russell, S. J., & Norvig, P. (1995). *Artificial intelligence: A modern approach*. Prentice Hall.
Schroder, H. M., Driver, M. J., & Streufert, S. (1967). *Human information processing: Individuals and groups functioning in complex social situations*. Holt: Rinehart and Winston.
Shneiderman, B., & Plaisant, C. (1987). *Designing the user interface: Strategies for effective human-computer interaction*. Readings: Addison-Wesley Publishing Company.
Simon, H. A. (1955). A behavioural model of rationality choice. *Quarterly Journal of Economics, 69*(1), 99–118.
Swar, B., Hameed, T., & Reychav, I. (2017). Information overload, psychological ill-being, and behavioral intention to continue online healthcare information search. *Computers in Human Behavior, 70*, 416–425.
Whittaker, S., & Sidner, C. (1996). *Email overload: Exploring personal information management of email.* Paper presented at the proceedings of the SIGCHI conference on human factors in computing systems, New York.
Woolridge, M., & Jennings, N. R. (1995). Intelligent agents: Theory and practice. *The Knowledge Engineering Review, 10*, 2.

The manufacturer's authorised representative in the EU is Springer Nature Customer Service Centre GmbH, Europaplatz 3, 69115 Heidelberg, Germany. If you have any concerns regarding our products, please contact ProductSafety@springernature.com

Printed and bound by CPI Group (UK) Ltd, Croydon, CR0 4YY
30/09/2026
02225695-0001